Seed Dormancy and Germination

Edited by Jose Carlos Jimenez-Lopez

Published in London, United Kingdom

IntechOpen

Supporting open minds since 2005

Seed Dormancy and Germination
http://dx.doi.org/10.5772/intechopen.77864
Edited by Jose Carlos Jimenez-Lopez

Contributors
Pedro Pablo Gallego, Manuel Ayuso, Mariana Landín, Esther Barreal, Angel Jesús Matilla, Cecilia Vega, Jorge Quezada, Milagros Bueno, Jamal R. Qasem, Ahmed Al-obaidi, Sergio Rodríguez Rodríguez, Iramis Vargas Díaz, Asterio Hijuelo Ramírez, Frederique Loumeto Ipolo, Juan José Silva Pupo, Jorge Pérez Pérez, Yanexis Fonseca Carrasco, Daliannis Oliva Domínguez, Yarisbel Gómez Masjuan, Michel Baldoquín Hernández

First published in London, United Kingdom, 2020 by IntechOpen
IntechOpen is the global imprint of INTECHOPEN LIMITED, registered in England and Wales, registration number: 11086078, 7th floor, 10 Lower Thames Street, London, EC3R 6AF, United Kingdom
Printed in Croatia

British Library Cataloguing-in-Publication Data
A catalogue record for this book is available from the British Library

Additional hard and PDF copies can be obtained from orders@intechopen.com

Seed Dormancy and Germination
Edited by Jose Carlos Jimenez-Lopez
p. cm.
Print ISBN 978-1-78984-860-1
Online ISBN 978-1-78984-861-8
eBook (PDF) ISBN 978-1-83968-454-8

For EU product safety concerns:
IN TECH d.o.o., Prolaz Marije Krucifikse Kozulić 3, 51000 Rijeka, Croatia, info@intechopen.com or visit our website at intechopen.com.

Meet the editor

Dr. Jose C. Jimenez-Lopez finished his PhD degree in Plant Science (2008) at the Spanish National Research Council (CSIC). Soon after, he moved to the USA to start his postdoctoral research at Purdue University, where in 2011 he worked on the biochemical and cellular characterization of actin-binding proteins involved in plant actin cytoskeleton dynamics and signaling. In 2012, he was awarded an EU Marie Curie research grant to work at the University of Western Australia. Currently, Dr. Jimenez-Lopez is a senior research fellow (Ramon y Cajal research program) working at CSIC on the functionality, health benefits, and allergy implications of proteins from reproductive tissues (pollen and seeds) in crop species (mainly legumes) of agro-industrial interest.

Contents

Preface

Different recent studies have discovered new exciting knowledge concerning seed dormancy and germination. Among the most significant discoveries are probably the epigenetic mechanisms controlling seed dormancy entry and release and germination events. While bioinformatics and systems biology have been employed in generating new hypotheses, experimental biology is being used to test them using different molecular tools such as the characterization of "seed dormancy mutants," the research outcomes if which look very conclusive and promising. More progress in seed dormancy and germination research will be necessary to reach new physiological understandings. Probing new hypotheses and using forward genetics and biochemical and molecular approaches seem to be the chosen ways for exploring emerging mechanisms to provide an overview of the frontier of this field.

Seed dormancy has played a significant role in the adaptation and evolution of seed plants. Its biological significance is clear; however, much more progress has to be made regarding the molecular mechanisms underlying seed dormancy induction, maintenance, and alleviation, which still remain elusive. The hormonal contribution to these processes, i.e. gibberellin and abscisic acid metabolism in seeds, has greatly added substantial insights into the current understanding of seed dormancy.

Interestingly, other mechanisms such as chromatin remodeling through histone modification, i.e. ubiquitination, methylation, and acetylation, could lead to transcription activation or gene silencing, critical for seed dormancy regulation. Furthermore, small interfering RNA and/or long non-coding RNA might be a trigger of epigenetic changes at the seed dormancy or germination stages.

Therefore, appropriate distribution of seed germination, in both temporal and spatial fashion, is critical for the survival and propagation of seed plants. Spatial distribution of germination is usually controlled through seed and fruit morphology, which improves seed offspring spreading. In contrast, temporal distribution of germination is mostly controlled by the physiological conditions of seeds. A difference in the physiological status among individual seeds in a population allows each seed to germinate at a different timing, which is an important strategy for seeds to avoid different backward situations such as competition with their relatives, while abolishing extinction of all individuals due to unsuitable environmental conditions. Plants have evolved acquiring seed dormancy and temporal suppression of germination under conditions favorable to germination. Induction of seed dormancy during the maturation stage and its release in a dry state after a certain period of time ("after-ripening") are extensive phenomena that can be observed in several species of seed plants.

There may be a universal mechanism of seed dormancy as well as a species-specific variation in the regulatory mechanisms. The progress in basic and applied seed science will enable knowledge translation, another frontier of research to be expanded for food production improvements with lower inputs in agriculture.

Dr. Jose Carlos Jimenez-Lopez
Spanish National Research Council (CSIC),
Spain

Section 1

Dormancy Process

Chapter 1

Analysis of the Effect of Scarification Process on Papaya (*Carica papaya* Lin.) Seeds Germination

Sergio Rodríguez, Iramis Vargas, Asterio Hijuelo, Frederique Loumeto, Juan J. Silva, Jorge Pérez, Quirino Arias, Yanexis Fonseca, Yarisbel Gómez, Michel Baldoquín and Daliannis Oliva

Abstract

The presence of the aril (sarcotesta) in the papaya causes a slow and low germination, being necessary to break the state of dormancy. Calcium hydroxide that was applied in order to evaluate its scarifying effect was the objective. The sample consisted of 60 randomly selected fruits of hermaphrodite plants in a commercial production batch of approximately 1 ha (2222 plants) showing commercial maturity, of homogeneous size (±2 kg). The treatments were calcium hydroxide $Ca(OH)_2$ at three doses, dipping the seed for a period of 24 h; the standardized sodium hydroxide (NaOH) at 25% with a 15 min immersion time. The highest germination and vigor seeds were obtained applied $Ca(OH)_2$ with highly significant differences respect to the rest treatment, especially for the dose of 60 g l^{-1} of water for reasons of diminishing expenses. Significant correlations were found, with direct relations (aril and mechanical damage) and inverse relations (abnormal seeds) between the variables evaluated related to the vigor and germination of the seeds. It can be an ecological and not expensive methodology to improve the germination and vigor of papaya seeds in relation to other chemical compounds to scarify.

Keywords: papaya, dormancy, seed germination, scarification, vigor, aril

1. Introduction

Fruit trees play an important role in human nutrition; and among these highlights, the papaya, this is a crop of tropical climate, widely appreciated for being one of the few fruit that provide continuous production throughout the year after the start of fruiting, to possess fruits with a high nutritional value, and to achieve high yields; generating good income to the families dedicated to their cultivation, due to the high prices that reaches in the market [1].

Carica papaya Linn belonging to family Caricaceae is known as papaya in France, United Kingdom, Mexico, Cuba, etc., papita in India, tree melon in Holland, paw paw in Australia and United Kingdom, and mamao in Brazil. The plant is native of tropical America. The properties of papaya fruit and other parts of the plant are also well known in traditional system of medicine. During the last few decades, considerable progress has been achieved regarding the biological activity and medicinal application of papaya and now it is considered as valuable nutraceutical fruit plant [2].

The world production of papaya occurs in more than 60 countries, according to FAO, for the year 2010, a production of 11,568,346 tons of the fruit was registered, with the main producing countries of the highest to lowest volume: India, Brazil, Indonesia, Nigeria, Mexico, Ethiopia, Colombia, Thailand, and Guatemala [3].

As a crop is an important source of employment, has a good yield, earliness to enter into production, and guarantees staggered crops throughout the year. Despite all these advantages, it does not reach the maximum productive potential, this is due in large measure to the problematic that it manifests in terms of the quality of seed [4]. Is consider one of the tropical fruits more appreciated for fresh consumption and for industrialization [5]. In Mexico like in Cuba, Maradol variety is the more cultivated, is a Cuban variety.

Different varieties of papaya are commercially propagated through seed an easy management and low cost, without taking into account the heterogeneity caused by crossed polinization [5].

In the sowing of papaya, it is best to use freshly harvested seeds, because while increasing the storage time of the seed, the germination rate decreases [6]. The majority part of the papaya sowing is done with stored dry seeds, and in this condition, the seed germination is erratic, asynchronous, slow, and incomplete [7], which diminishes the germination percent [5]. The desiccation produces stress in the papaya seeds when the moisture content lowers to 8.0% [8], being the cause of the seed dormancy or metabolic quiescence [9, 10]. In Taiwan, the papaya industry is limited by seed germination rates [11]. This is attributed to the presence of inhibitors as the phenolic compounds in the sarcotesta and seed coat [12, 13], and in some cases, the seeds lack embryos [6].

The seed is enclosed in a gelatinous sarcotesta (aril or seed coat), which is formed from the outer integument [14]. The sarcotesta can delay germination, and also dormancy is observed in seeds from which the aril has been removed [15].

Papaya, like many plants, presents as one of its main problems in its reproduction, the dormancy of the seed which influences the quality of it; because by reaching its maximum point of maturity, it initiates a period of latency produced by internal and external factors. It is normally interrupted when the natural conditions suitable for germination are present or when treatments are used that help to propitiate these ideal conditions and increase the percentages of germination [16], but in vitro conditions favored germination of papaya more than in vivo environment [17].

The seed of papaya is characterized by being bitegumented, since the internal tegument originates the tegumen and the external one to the testa, which is multiplicative up to 60 layers and 3 distinctive strata: endotesta, mesotesta, and exotesta (sarcotesta). This last one of semipermeable consistency, high humidity and concentrate phenolic compounds that, as a whole, induce latency. This causes the inhibition of fluid and gas exchange, delayed dehydration and colonization of pathogenic microorganisms. In [18] are mentioned others researchers who investigated this problem [19, 20].

One of the ways to break the latency of the seeds and make them have a good quality overall, increasing the percentages of germination is using the different methods of scarification.

The methods of scarification include physical, mechanical, and biological treatments such as dry heat, the rupture of the testa, the soaking in water, and chemical solutions that promote the germination of the seeds, where any treatment that destroys or reduces the impermeability is called scarification, so in some cases, it is only enough to destroy a single point of the cover to produce the imbibitions and exchange of gases and thus initiate the germination [21].

Apparently, latency is a survival mechanism in the presence of certain climatic conditions: very low temperatures, alternations of dry and humid times, and desert climates. The exact causes of the latency phenomenon are unknown, and on the other hand, when the latency is due to testa conditions, the lethargy ends at the moment that it cracks or weakens by mechanical or chemical actions or by effect of the environment [22, 23].

Different seed treatments to promote germination and to reduce germination time are mentioned in [24], sowing seeds and at warm temperatures, exposing dry seeds to 10°C prior to sowing, drying seeds and soaking seeds in distilled water, potassium nitrate, thiourea, sodium thiosulfate, tannic acid or ferulic acid. The same authors described contradictory results using gibberellins on papaya seed germination. But in your research, they demonstrated that dehydration to 5.3% or 6.9% and 6.8% moisture content, followed by exposure to subzero temperatures and treatment with GA3, were the most favorable combined treatments to enhance papaya seed germination.

The used of smoke water on seed germination and seedling growth of papaya, cultivar Tainung No. 2 consistently and significantly increased the percentage of nitrogen in roots and shoots and significantly increased the percentage of magnesium in shoots. In these experiments, smoke-water showed potent germination promotion at low concentrations and promoted multiple growth attributes such as chlorophyll content and seedling vigor index at all concentrations in papaya seedling production [11].

Pregerminative treatments are used to break the latency status of the seeds. In [25] are mentioned stratification and scarification. Scarification is any process that breaks, scratches, mechanically alters or softens the covers of the seeds to make them permeable to water and gases.

Seed scarification methods have been developed and modified over time to make these more practical and effective. Important methods of seed scarification include heat, freeze–thaw, mechanical, and acid scarification [16].

In the scientific literature, some types of scarification are described, such as mechanical [26–28], physical [29–31], chemical [32, 33], and biological. Mechanical, physical, and biological scarification have disadvantages in relation to chemical scarification, because they require more time, are laborious and inadequate to condition large quantities of seed; while chemical scarification still requires more research [18], especially with calcium hydroxide.

There are chemical substances used to scarify seeds, among the most used are the sulfuric acid [34], sodium hydroxide, and hydrochloric acid [7, 18, 20, 31]. The positive effect of the use of NaOH in the benefit of papaya seed is that simulates natural degradation of sarcotesta and improves the conditions of the seed, so it is a viable alternative for use in conditioning seed [18]. Other results have been demonstrated that the combination of NaOH treatment and stratification is an effective practice to break *Iris lactea* var. chinensis seeds dormancy and improve germination percentage [35].

Chemistry scarification is considered as one of the most effective scarification methods used for seed scarification. Sulfur acid is the most popular and effective chemical product for acid scarification. The effectiveness of acid scarification depends on concentration of acid duration of scarification and species and cultivars used [16, 36].

Traditionally, it has been used to separate the mucilage from the papaya ferment the seeds in water at different time intervals and the sunny one for 2 or 3 h [37].

The objective of the research was to evaluate the scarifying effect of calcium hydroxide on the germination and vigor quality of papaya seeds, Maradol variety.

2. Experiment location

The experiment was carried out in the Laboratory of Seed Test of the plant of Benefit Manuel Espinosa Ramírez of the business unit of seeds base Granma, belonging to the company producer and marketer of seeds, using seed of papaya 'Red Maradol', collected in areas of the Experimental Station Jucaibama of the Agricultural Research Institute Jorge Dimitrov, Bayamo, Granma province.

2.1 Development of the experiment

The sample consisted of 60 randomly selected fruits of hermaphroditic plants in a commercial production lot of approximately 1 ha (2222 plants), showing commercial maturity (two strips), of homogeneous size (±2 kg). The seeds were extracted, and the batch was homogenized; 200 g of fresh dough were deposited with 500 ml plastic flasks representing each experimental unit.

2.2 Treatments

The treatments were composed of the solution of calcium hydroxide ($CaOH_2$) at three doses (60, 80, and 120 g l^{-1} of water) by dipping the seed for a period of 24 h, the standardized sodium hydroxide (NaOH) at 25% with a 15-min immersion time, for a total of four treatments plus the control and six replicates. The control consisted in fermenting in running water the seed for 24 h.

To eliminate the sarcotesta (aril), the seeds were rubbed between two jute cloths where the time needed to remove all the aril of the material was evaluated, being the optimal time to use of 15 s for each treatment given the amount of material to process. Immediately, they were rinsed three times with running water, spread over a sieve in the shade and at room temperature (28 ± 1°C) for drying for 48 h.

The physical quality of the seed was determined by the effectiveness of the product, physical appearance, and mechanical damage within 3 days of the treatment, compared to the control; the physiological quality was determined by the germination percentage 7 days after sowing (vigor), being valued by the germination rate [38], and 28 days for final germination (as indicated in the germination standard, with the method in sand). The seed was soaked for a term of 24 h and placed in previously disinfested aluminum trays at a temperature of 100°C.

The incubation was carried out in the germinating chamber Paul Polikeit, Model HALLE S. A, with 80% of relative humidity, 40 ± 2°C of temperature and natural light; for the sanitary quality, it was determined by assembling all the treatments with the method between paper (BP), evaluating by observing the evidences of the development of microflora on the seed, during the germination test.

To carry out this test, three repetitions of 25 seeds were used by extraction, it was put to incubated in water the seeds for a time of 24 h, after the time elapsed, each seed was sectioned in longitudinal form leaving the cotyledons visible placed in culture tubes wrapped with aluminum foil adding a solution of 2, 3.5-triphenyl chloride tetrazolium to 1%, and the tubes were placed in an incubator at 35 ± 1°C for a time of 2 h.

2.3 Experimental design and evaluated variables

The experimental design used was completely randomized with bifactorial arrangement and six replicates. The variables assessed were vigor (vigor, 120, 240, and 360 days of conservation at 4–8°C, in percent); germination (germination, 120, 240, and 360 days of conservation at 4–8°C, in percent); the time needed to eliminate aril (s), mechanical damage, MD (%), and abnormal plants, AP (%), according to ISTA Methodology [38].

The data for each measured variable were statistically processed to check compliance with the normal distribution of the data (Kolmogorov-Smirnov test) and the homogeneity of the variances (Bartlett test). These two premises of the analysis of variance were not met, even after testing several data transformation equations, so we proceeded to the application of nonparametric variance analysis through Kruskal-Wallis, to demonstrate the existence or not of variability between treatments with a probability level of 0.05. The averages (aver.) of the treatments, the standard deviation (sd) of the mean, and the significance are shown. The multiple comparisons between treatments were made through the differences between the averages of the ranges [39].

We also performed analysis of partial correlations between variables with the use of Spearman correlating coefficient, with the aim of determining the existence of linear relationship between selected variables. Those variables that could have a direct or inverse relationship were selected in relation to the vigor and germination in their different times used as the aril, the mechanical damages, and the percentage of abnormal seeds.

Statistical processing was carried out with the use of statistical packages MINITAB 13 [40] for the test of homogeneity of variances and Infostat 2017 [41] for the rest of the statistical analyses.

3. Results and discussion

Significant differences were found between the different treatments in the vigor of the seeds (**Table 1**). The highest level of vigor of the seed was due to the use of calcium hydroxide, more than sodium hydroxide and fermentation. This tendency was maintained during the different storage times of the seeds at a constant temperature. Significant although not shown statistically, there is evidence of a decrease in

Vigor (%)					
Compound	**Dose**	**0 days**	**120 days**	**240 days**	**360 days**
		Aver. ± SD	**Aver. ± SD**	**Aver. ± SD**	**Aver. ± SD**
$Ca(OH)_2$	60 g l^{-1}	80.7^{ab} ± 2.8	83.7^{a} ± 2.7	77.3^{b} ± 2.3	71.7^{ab} ± 5.7
$Ca(OH)_2$	80 g l^{-1}	81.7^{ab} ± 1.6	81.2^{ab} ± 1.2	79.3^{ab} ± 2.3	60.0^{bc} ± 6.8
$Ca(OH)_2$	100 g l^{-1}	89.3^{a} ± 1.2	82.3^{a} ± 2.2	82.3^{a} ± 1.7	75.7^{a} ± 3.6
NaOH	25 g l^{-1}	50.3^{c} ± 4.2	42.7^{c} ± 3.7	26.5^{c} ± 3.4	6.7^{c} ± 0.5
Fermentation		75.3^{bc} ± 1.9	75.8^{bc} ± 3.4	73.5^{bc} ± 3.5	71.7^{ab} ± 4.8

Different letters indicate significant differences to $p \leq 0.05$ through the differences between the average of the ranges.

Table 1.
Effect of calcium hydroxide and sodium hydroxide with different doses on the vigor of the seeds in different storage times.

Germination (%)					
Compound	Dose	0 days	120 days	240 days	360 days
		Aver. ± SD	Aver. ± SD	Aver. ± SD	Aver. ± SD
$Ca(OH)_2$	60 g l^{-1}	88.0^{ab} ± 2.18	87.3^{b} ± 1.6	80.2^{bc} ± 1.73	89.3^{ab} ± 6.6
$Ca(OH)_2$	80 g l^{-1}	89.0^{ab} ± 1.9	89.0^{ab} ± 1.6	81.8^{ab} ± 1.2	70.5^{bc} ± 8.2
$Ca(OH)_2$	100 g l^{-1}	94.0^{a} ± 2.9	92.3^{a} ± 2.0	85.3^{a} ± 1.8	88.8^{a} ± 4.3
NaOH	25 g l^{-1}	65.3^{c} ± 4.2	56.0^{c} ± 3.2	42.0^{d} ± 3.0	15.5^{c} ± 3.1
Fermentation		83.7^{bc} ± 2.3	83.7^{bc} ± 2.7	76.3^{cd} ± 2.9	84.3^{ab} ± 6.08

Different letters indicate significant differences to $p \le 0.05$ through the differences between the average of the ranges.

Table 2.
Effect of calcium hydroxide and sodium hydroxide with different doses in germination percentage of seeds.

the seeds vigor for all the evaluated treatments by increasing the conservation time of papaya seeds, which suggests that it is more efficient to apply calcium hydroxide in order to improve the response of papaya seeds with a minimum storage time.

The variability in the response of calcium hydroxide could be a consequence of the fact that some seeds within the same batch have a more persistent dormancy than others and that small and large seeds can be found in the same batch [42].

Sodium hydroxide reached lower percentages, even less than 60%, which is the minimum value established for Cuba for this crop [43, 44].

The physiological quality of the papaya seed is characterized by a high sensitivity to several factors with respect to germination and vigor, considering that there are integrating elements of great importance at the plantation level [45].

For the germination (**Table 2**), the application of calcium hydroxide obtained the best results, with percentages over the 80% in comparison with the rest of the treatments. Germination was more affected when the seeds were treated with sodium hydroxide and to the extent that the storage time of the seed was increased. Like the vigor, the germination percentage decreases in the treatments when the conservation time increased.

Physiologically, these results could be interpreted as a sequence of events of deterioration that begins with problems of functionality in the seminal membranes, which causes an excessive flow of cellular constituents, evidenced this by the high absorbance values and consequent loss of metabolites, the magnitude of which can restrict the germinative process [46].

Cold stored papaya seeds maintained significantly higher germination and better seedling vigor than the room stored seeds. With the increase in the duration of storage seed germination decreased after 20 mo. at room temperature, it declined marginally during the same period when kept in cold storage. Irrespective of the storage conditions, seeds kept in sealed polythene bags or plastic bottles had better germination and seedling vigor than those on paper and cloth bags. Shoot length and dry weight decreased significantly with the increase in the duration of storage. Viability of papaya seeds can be maintained considerably at room temperature up to 8 mo. by storing the seed in sealed, preferably airtight, polythene bags or plastic bottles. Cold storage using polythene bags or plastic bottle is recommended [47].

The storage conditions are very important. According to the classification of seed storage behavior, the papaya seed is classified as recalcitrant seed [48], others in intermediate seed [49]. In storage behavior ambient conditions, the papaya seeds survive for a short period of time [50] and are considered intermediate between recalcitrant and orthodox attribute and deteriorate rapidly at higher storage temperatures and relative humidity. Fresh seeds give higher germination rate and seedling vigor that will decline with increasing the storage time [51] and consider that the best conditions for papaya seed storage is when containing 6.0% moisture and stored at 0°C gave higher percentage of germination, lower dormancy, and seed death.

In [18], the treatment that most affected the germination was the application of sodium hydroxide and the higher incidence of microorganisms, with high percentages of plants affected by fungi, which remained even below the germination approved for marketing, which requires, more than 60% [43].

In [52], described some pre-sowing treatments with the finality of increase the germination, like preconditioned papaya seed at 24°C before transfer to 32°C, soaked in KNO_3 for 30 min, soaking in gibberellic acid (GA_3; 200 ppm) for 24 h resulted in highest germination percentage in soil compared to α-naphthalene acetic acid (NAA) and KNO_3 treatment, and using a protocol of sterilization and germination of papaya seeds in response to light emitting diodes and got 100% of sterilization and 100% of germination.

The efficiency of the different scarification methods has been demonstrated in several investigations to favor the germinative process. In [53] were evaluated several methods of scarification of pacain seeds (*Chamaedorea* sp.), where the germination results indicated that the highest percentage was 77.8% and that corresponded to cold water treatment for 15 days (physical scarification), followed by treatment with mechanical scarification by hammer blow with 77.17% compared to the absolute control that obtained a 5.67% of germination.

With chemical scarification methods also obtained positive results [54], and verified germination percentages of 97 and 94%, respectively, in seeds of *Centrosema macrocarpum* for 10 min immersion.

The treatments used release the aril at different time intervals and there are differences between them in relation to the mechanical damage and the quantity of abnormal seeds of the Maradol variety (**Table 3**). The best results were obtained when calcium hydroxide was applied at the 100 l^{-1} dose and with sodium hydroxide. Those compounds only needed an average time of 12 s to achieve an efficiency in the detachment of the aril, while with the application of the control dose, the worst results were obtained, with an average of 38 s that in some cases reached reaching more than 420 s to achieve the release of the aril, so that with this result, it is inferred that to work 25 kg of wet seeds, it would take approximately 38 min if treated with sodium hydroxide or calcium hydroxide in 100 l dose and approximately 106 min for the case fermented only with water.

Compound	**Dose**	**Aril (s)**	**MD (%)**	**AP (%)**
		Aver. ± SD	**Aver. ± SD**	**Aver. ± SD**
$Ca(OH)_2$	60 $g\,l^{-1}$	21.2^{cd} ± 1.3	1.0^{a} ± 0.9	1.5^{ab} ± 1.5
$Ca(OH)_2$	80 $g\,l^{-1}$	17.7^{cd} ± 2.5	1.0^{a} ± 0.9	0.7a ± 1.0
$Ca(OH)_2$	100 $g\,l^{-1}$	13.3^{ab} ± 1.4	0.5^{a} ± 0.6	0.4^{a} ± 0.7
NaOH	25 $g\,l^{-1}$	12.0^{a} ± 0.9	0.5^{a} ± 0.2	3.3^{b} ± 1.0
Fermentation		37.7^{d} ± 6.3	9.2^{b} ± 1.2	0.8^{a} ± 0.9

Different letters indicate significant differences to $p \leq 0.05$ through the differences between the average of the ranges.

Table 3.
Effect of calcium hydroxide and sodium hydroxide with different doses in the time (s) of detachment of the seed's aril.

When carrying out an essay [55], with different methods to improve the germination of four forage shrubs legumes [Tagasaste (*Cytisus proliferus*) and three species of Teline (Genista)], obtained the most encouraging results for the elimination of the aril in the treatment with concentrated sulfuric acid for 30 min, demonstrating the effectiveness of chemical compounds in the scarification of seeds.

Refs. [7, 20] observed a negative effect on the papaya germination due to the presence in the aril or sarcotesta of inhibitory substances. It is also said that the marked decrease in germination, in the presence of sarcotesta is due to the low oxygenation of the seeds, which is why it is recommended to remove it. Likewise, in a study carried out in Honduras with seeds of the papaya, Maradol variety, higher percentages of germination were obtained with freshly harvested and oared seeds (83%), while fresh seeds with burned arils showed a lower percentage of germination, with 75% [56].

Similar results presented by [18, 20] which treated seeds with very corrosive products not only the germination is affected, but that several plants emerged with problems fundamentally in the radicle and hypocotyl, assuming the influence of other factors such as fluctuation in the moisture of the seed in the time of conservation.

The variables that were selected for the multiple correlation analyses (**Table 4**) showed that the release of aril seed correlated significantly with germination and vigor 360 days, positively and in a mean size. No correlations were found between this variable and the rest of the values of germination and vigor evaluated. Candiani et al. [57] concluded that the germination of *Michelia champaca* L. seeds is hindered probably due to the presence of inhibiting substances in the aril, is considered as endogenous causes of seed dormancy which include factors such as phytohormones, or by interference with water uptake [58]. Abscisic acid (ABA) is the key inhibitor of germination in *Taxus yunnanensis* seeds during wet sand storage [59].

The mechanical damage caused to the seeds during this process in this experiment did not correlate with the different levels of germination and vigor that were studied; however, the percentages of abnormal plants in the different time intervals evaluated showed a significant correlation, with medium to high values, but inverse and indicates that as the percentage of abnormal plants increases in a seed lot, the number of seeds germinates decreases and the vigor. Germination vigor is driven by the ability of the plant embryo, embedded within the seed, to resume its metabolic activity in a coordinated and sequential manner.

Correlations analysis

	Aril	**MD**	**Adnor**
Ger 0	0.13 ns	0.07 ns	−0.74*
Ger120	0.24 ns	0.15 ns	−0.71*
Ger240	0.25 ns	0.18 ns	−0.72*
Ger360	0.46*	0.34 ns	−0.70*
Vigor	0.13 ns	0.08 ns	−0.73*
Vigor120	0.28 ns	0.19 ns	−0.70*
Vigor240	0.30 ns	0.22 ns	−0.69*
Vigor360	0.46*	0.34 ns	−0.70*

Ger is germination, MD is mechanical damage, and adnor is adnormal plants.
**Indicate significant differences to $p \leq 0.005$.*

Table 4.
Spearman correlations coefficient between variables.

Was analyzed the vigor tests on lettuce (*Lactuca sativa* L.) seeds and their correlation with emergence [60], demonstrated that saturated salt accelerated aging and digital image analysis were the best laboratory tests for lettuce seed vigor evaluation, especially for seed lots to be used for plug seedling production. In some case, the use of seed vigor tests is used to predict field emergence in plants, like lucerne (*Medicago sativa*) [61].

In studies [62] about studied the correlations of seed germination percent of two sweet corn hybrids (*Zea mays* L.) with field emergence and some measured traits related to yield, the results of correlation analysis indicated that there was a high positive correlation between seed germination ability and vigor with seedling field emergence and most of the measured traits, as the percent of the radicle emergence.

The germination vigor depends on multiple biochemical and molecular variables. Their characterization is expected to deliver new markers of seed quality that can be used in breeding programs and/or in biotechnological approaches to improve crop yields [63].

Calcium hydroxide has great potential to be used as biocide in agriculture, because it has the advantage of not being phytotoxic, is economic and easy to use and is harmless to the environment and to humans. Decreased Ca levels in the nutrient medium reduced soybean leaf dry matter during seed fill, seed production, seed Ca concentration, and seed germination and increased the incidence of seedling disorders such as watery hypocotyl and epicotyl necrosis [64].

In [65] performed standard germination tests, germination and growth rate, accelerated aging, electrical conductivity, respiration rate and ATP content, to evaluate the vigor of the seeds of *Bromus biebersteinii* Roem & Schult and determined that the correlations with the behavior in the field of that forage were accelerated aging, respiration rate, and ATP content. The cited authors conclude that a vigor test alone is not adequate to measure the quality of seed lots. A combination of tests which measure both physiological and biochemical aspects should be used.

Some researches [66] showed that seed size is an important factor for germination and seedling vigor, establishing that larger seeds produce more vigorous seedlings but with slower emergence. So it can be argued that by using the largest scarified seeds that we have, we can help considerably to decrease the percent of abnormal plants and at the same time increase the vigor and the percentage of germination, variables evaluated in the present investigation.

To correlate characters related to seed germination, is important to investigate the effects of environmental factors prevailing during seed maturation under controlled conditions to understand exact reasons for unusual seed dormancy and germination requirements, for example, the germination of *Citrullus colocynthis* (*Cucurbitaceae*) is very sensitive to light and incubation temperature as well as to the environmental conditions associated with the time of seed maturation [67] and seed dormancy is a temporary failure of a viable seed to complete germination under normally favorable physical environmental conditions [33].

Seed germination tests assess the ability of the seed to produce a healthy plant when placed under favorable environmental conditions. Germination tests are conducted for a prescribed time period under laboratory conditions that assure optimum moisture, temperature, and light. Unfortunately, these conditions are seldom encountered in the field, and field emergence may be overestimated by standard germination tests. Seed lots that have low germination also are less vigorous due to seed deterioration. As seeds deteriorate, loss of vigor precedes loss of viability, so seeds with low germination usually will be less vigorous. Hence, in seed lots with poor germination, those seeds that do germinate often produce weaker seedlings with reduced yield potential. However, some species (such as many native grasses) have inherently low germination potential and cannot be assumed to have poor vigor due to low germination. Seed vigor usually cannot be assessed by the consumer. Germination is a good indicator of seed vigor [63, 68].

Using efficient methods to scarify papaya seeds can increase the germination percentages of seeds, only if there is good control of environmental factors, because papaya seed germination is affected by light, temperature, oxygen, pH, and the moisture of the substrate [7].

It is necessary to conduct researches about the biometric and morphological characteristics of fruit and seeds, aiming at the maximum germination capacity and seed vigor [69], because biometric studies of seeds and their phenotypical correlations allow the quantitative evaluation of a character's relevance in relation to another [70]. It continues to investigate the correlations between the different indicators that can characterize the quality of the seeds. In adaptive correlations between seed size and germination time [71], present a model for the coevolution of seed size and germination time within a season when both affect the ability of the seedlings to compete for space and show that even in the absence of a morphological or physiological constraint between the two traits, a correlation between seed size and germination time is nevertheless likely to evolve.

Seed germination is a complex process and we need to understand the underlying molecular, hormonal, and mechanical aspects [72]. The environment during seed production has major impacts on the behavior of progeny seeds [73]. For that reason, the seed biology is considered the principal research topic for food security take into consideration the climate change [72].

Nowadays, there are advances in the propagation of papaya by biotechnological methods. Efficient micropropagation of papaya has become crucial for the multiplication of specific sex types of papaya and in the application of genetic transformation technologies. Significant progress has been achieved using organogenesis and somatic embryogenesis as the shoot tip, axillary bud and single node culture, organogenesis, anther and ovule culture, and regeneration from protoplasts, callus induction and somatic embryogenesis and the mass propagation by ex vitro rooting and acclimatation [74].

4. Conclusions

In natural conditions, the germination of papaya seeds has difficult by the presence of aril (sarcotesta) that become in a physical barrier which limits the diffusion of water and gases into the seeds and by the effect of phytohormones which preventing germination of seeds, causing dormancy, limiting the development of the embryos and causes a low and variable germination affecting the final percentage. This problem can be solved with the scarification of papaya seed. The results of this research showed that the dormancy by the presence of aril is produced in the papaya seed can be broken with the use of NaOH; but higher results were achieved with the use of calcium hydroxide, $Ca(OH)_2$. The results suggested that chemical scarification with calcium hydroxide can improve germination percentage and vigor of the papaya seeds, take into account that the seed is considered the major essential input in crop production. If the seed quality parameters (vigor, germination rate) decrease, the yields are affected. The scarification of papaya seeds with the use of calcium hydroxide for its proven effect in this research on the benefit of papaya seeds and easy acquisition and will reduce the costs of seed. But a good germination of the papaya seedlings depends on the many environmental factors and excellent agronomy practice. Finally, the effectiveness of scarification methods could change among cultivars within the papaya specie.

Author details

Sergio Rodríguez[1*], Iramis Vargas[2], Asterio Hijuelo[3], Frederique Loumeto[4], Juan J. Silva[1], Jorge Pérez[1], Quirino Arias[5], Yanexis Fonseca[1], Yarisbel Gómez[3], Michel Baldoquín[3], Daliannis Oliva[3]

1 Plant Biotechnology Study Center, Faculty of Agricultural Science, University of Granma, Cuba

2 Seed Production and Marketer Enterprise, Ministry of Agriculture, Bayamo, Granma, Cuba

3 Faculty of Agricultural Science, University of Granma, Cuba

4 Educagri, Ministry of Agriculture and Food, French, Guiana

5 Applied Chemistry Study Center, Faculty of Engineering, University of Granma, Cuba

*Address all correspondence to: srodriguezr@udg.co.cu

References

[1] Gallardo D. Análisis de la comercialización de la papaya variedad Maradol (*Carica papaya*) producida en el municipio de Tapachula, Chiapas para exportación al mercado estadounidense [Internet]. 2016. Available from: http://repositorio.uaaan.mx:8080/xmlui/handle/123456789/8076 [Accessed: 2019-01-18]

[2] Yogiraj V, Kumar P, Singh C, Goyal A, Vyas B. *Carica papaya* Linn: An overview. International Journal of Herbal Medicine. 2014;**2**:01-08

[3] Food and Agriculture Organization of the United Nations (FAO). FAOSTAT database. 2016

[4] Peña I. Enfermedades virales en el cultivo del papayo (*Carica papaya L.*). Revista CitriFrut. 2008;**25**:13-23

[5] Andrade M, Ayala J, Alia I, Rodríguez H, Acosta C, López V. Effect of germination promoters and substrates in the development of papayo seedlings. Revista de la Facultad de Agronomía—LUZ. 2008;**25**:617-635

[6] Nagao MA, Furutani SC. Improving germination of papaya seed by density separation, potassium nitrate, and gibberellic acid. HortScience. 1986;**21**(6):1439-1440

[7] Lange A. Effect of the sarcotesta on germination of *Carica Papaya*. Botanical Gazette. 1961;**122**:305-310

[8] Ellis RH, Hong TD, Roberts EH. Effect of storage temperature and moisture on the germination of papaya seeds. Seed Science Research. 1991;**1**:69-72

[9] Magill W, Deighton N, Pritchard HW, Benson EE, Goodman BA. Physiological and biochemical studies of seed storage parameters in *Carica papaya*. Proceedings of the Royal Society of Edinburgh. 1994;**102B**:439-442

[10] Wood CB, Pritchard HW, Amritphale D. Desiccation induced dormancy in papaya (*Carica papaya* L.) seeds is alleviated by heat shock. Seed Science Research. 2000;**10**:135-145

[11] Chumpookam J, Huey L, Ching S. Effect of smoke-water on seed germination and seedling growth of papaya (*Carica papaya* cv. Tainung No. 2). HortScience. 2012;**47**(6):741-744

[12] Reyes MN, Pérez A, Cuevas J. Detecting endogenous growth regulators on the sarcotesta, endosperm and embryo by paper chromatography on fresh and old seeds of two papaya varieties. Journal of Agriculture-University of Puerto Rico. 1980;**64**:164-172

[13] Chow YJ, Lin CH. *p*-Hydroxy benzoic acid as the major phenolic germination inhibitor of papaya seed. Seed Science and Technology. 1991;**19**:167-174

[14] Andreoli C, Khan A. Improving papaya seedling emergence by matriconditioning and gibberellin treatment. HortScience. 1993;**28**(7):708-709

[15] Yahiro M, Oryoji Y. Effects of gibberellins and cytokinin treatments on the promotion of germination in papaya, *Carica papaya* L., seeds. Memoirs of the Faculty of Agriculture, Kagoshima University. 1980;**16**:45-51

[16] Kimura E, Islam M. Seed scarification methods and their use in forage legumes. Research Journal of See Science. 2012;**5**:38-50. DOI: 10.3923/rjss.2012.38.50

[17] Bhattacharya J, Khuspe S. In vitro and in vivo germination of papaya (*Carica papaya* L.) seeds. Scientia Horticulturae. 2001;**91**:39-49. DOI: 10.1016/S0304-4238(01)00237-0

[18] Romero J, Mejía J, Carballo A, López A, Rangel J, Ávila C. Escarificación química de semilla de papaya. Revista Mexicana de Ciencias Agrícolas. 2013;**4**:947-954. DOI: 10.29312/remexca.v4i6.1162

[19] Kubitzki K. Caricaceae. In: Kubitzki K, Bayer C, editors. The Families and Genera of Vascular Plants. Vol. 5. Berlin: Springer-Verlag; 2003. pp. 57-61

[20] Tokuhisa D, Dias D, Alvarenga E, Dias L, Marin S. Tratamentos para superação da dormência em sementes de mamão. Revista Brasileira de Sementes. 2007;**29**:131-139

[21] Zare S, Tavili A, Javad M. Effects of different treatments on seed germination and breaking seed dormancy of *Prosopis koelziana* and *Prosopis juliflora*. Journal of Forestry Research. 2011;**22**:35-38. DOI: 10.1007/s11676-011-0121-8

[22] Sanabria D, Silva R, Oliveros M, Barrios R. Escarificación química y mecánica de semillas subterráneas de *Centrosema Rotundifolium*. Bioagro. 2001;**13**:117-124

[23] Varela SA, Arana V. Latencia y germinación de semillas. Tratamientos pregerminativos. In: Varela SA, Aparicio A, editors. Serie técnica: "Sistemas Forestales Integrados" Área Forestal—INTA EEA Bariloche Sección: "Silvicultura en vivero". 2017. pp. 3-6. Cuadernillo N° 3: ISSN: 1853-4775

[24] Salomão A, Mundim R. Germination of papaya seed in response to desiccation, exposure to subzero temperatures, and gibberellic acid. HortScience. 2000;**35**(5):904-906

[25] Dürr C, Dickie J, Yang X, Pritchard H. Ranges of critical temperature and water potential values for the germination of species worldwide: Contribution to a seed trait database. Agricultural and Forest Meteorology. 2015;**200**:222-232. DOI: 10.1016/j.agrformet.2014.09.024

[26] Goor AY, Barney CW. Forest Tree Planting in Arid Zones. 2nd ed. New York: The Ronald Press; 1976. 504 p

[27] Cabello A, Camelio M. Germinación de semillas de maitén (*Maytenusboaria*) y producción de plantas en vivero. Revista Ciencias Forestales. 1996;**1**:3-17

[28] Pérez A. Evaluación de doce métodos de escarificado en semillas de Chonte (*Zanthoxylum aguilarii*) y Canoj (*Ocoteaguate malensis*) en el Asintal, Retalhuleu [thesis]. Quetzaltenango, Guatemala: 2008

[29] Sánchez Y, Ramírez M. Tratamientos pregerminativos en semillas de *Leucaena leucocephala* (Lam.) de Wit. Y *Prosopisjuliflora* (Sw.) DC. Revista de La Facultad de Agronomía. 2006;**23**:257-272

[30] Coronel de Renolfi M, Cardona G, Ewens M, Ibarra E. Productividad y requerimientos de mano de obra en la producción comercial de plantines de algarrobo blanco (*Prosopis alba*). El caso del vivero forestal de la Estación Experimental Fernández, Santiago del Estero. Quebracho. 2013;**21**:26-38

[31] Jones C, Stevens M, Jolley V, Hopkins B, Jensen S, Turner D, et al. Evaluation of thermal, chemical, and mechanical seed scarification methods for 4 Great Basin lupine species. Native Plants Journal. 2016;**17**:5-17

[32] Vilela A, Ravetta D. The effect of seed scarification and soil-media on germination, growth, storage, and survival of seedlings of five species of *Prosopis* L. (Mimosaceae). Journal of Arid Environments. 2001;**48**:171-184. DOI: 10.1006/jare.2000.0735

[33] Baskin JM, Baskin CC. A classification system for seed dormancy.

Seed Science Research. 2004;**14**:1-16. DOI: 10.1079/SSR2003150

[34] Rostami A, Shavasar A. Effects of seed scarification on seed germination and early growth of olive seedlings. Journal of Bioloical Sciences. 2009;**9**:825-828

[35] Sun V, Zhang Y, Wang K, Qiu X. NaOH scarification ans stratification improve germination of *iris lacteal* var. chinensis seed. HortScience. 2006;**41**:773-774. DOI: 10.21273/HORTSCI.41.3.773

[36] Alderete A, de la Cruz N, Guerra J, Guevara E, Gelabert R. Promotion of germination of *Bauchinia divaricata* L. seeds by effects of chemical scarification. Research Journal of See Science. 2011;**4**:51-57. DOI: 10.3923/rjss.2011.51.57

[37] García JL, Vázquez ME, Torres MA, Dávila SI, Sánchez D. Métodos de extracción de semilla en papaya Golden y la relación con la longevidad. Revista Mexicana de Ciencias Agrícolas. 2011;**2**(2) Texcoco

[38] ISTA—International Seed Testing Association. International Rules for Seed Testing Edition 2005. Basserdorf, CH-Switzerland: International Seed Testing Association (ISTA); 2005

[39] Conover WJ. Practical Nonparametric Statistics. New York, USA: Wiley; 1999. 462 p

[40] Minitab Inc. Minitab Statistical Software, Version 13 to Windows. State College, Pennsylvania: EUA; 2009

[41] Balzarini MG, Gonzalez L, Tablada M, Casanoves F, Di Rienzo JA, Robledo CW. Infostat. Manual del Usuario. Córdoba, Argentina: Editorial Brujas; 2008

[42] Milberg P, Andersson L, Thompson K. Large-seeded species are less dependent on light for germination than small-seeded ones. Seed Science Research. 2000;**10**:99-104. DOI: 10.1017/S0960258500000118

[43] NRAG. Norma cubana para el cultivo No 120. Semillas de papaya. MINAG. Base legal de la dirección de semillas. República de Cuba: Ministerio de la Agricultura; 2011. pp. 83-93

[44] Ramos R, Hernández V, Ramos J. Tecnología para la Producción y Beneficio de Semillas de la Papaya (*Carica papaya L*). Empresa de Semillas. ACTAF. MINAG. La Habana: Editora Agroecológica; 2018. p. 35

[45] Serbelló F, Mesa J, Soto R. Efecto de diferentes alternativas biológicas, sobre el porcentaje y velocidad de germinación de las semillas de fruta bomba (*Carica papaya* L.). Agroecosistemas. 2014;**2**:247-253

[46] Bewley JD, Black AM. Seeds: Physiology of Development and Germination. 2nd ed. NY, USA: Plenuim Press; 1994. 445 p

[47] Singh RM, Singh ID. Effects of method and duration of storage on seed germination and seedling vigour in papaya (*Carica papaya* L.). Seed Research. 1981;**9**:67-72

[48] Hofmann P, Steiner AM. An updated list of recalcitrant seeds. Landwirtschaftliche Forschung. 1989;**42**:310-323

[49] Ellis RH, Hong TD, Roberts EH. An intermediary category of seed behaviour? 1. Coffee. Journal of Experimental Botany. 1990;**41**:1167-1174

[50] Ellis RH. Revised table of seed storage characteristics. Plant Genetic Resources Newsletter. 1984;**58**:16-33

[51] Zulhisyam AK, Tse C, Anwar A, Azwanida NN, Shazani S,

Jamaludin MH. Effect of storage temperature and seed moisture contents on papaya (*Carica papaya* l.) seed viability and germination. Journal of Sustainability Science and Management. 2013;**8**(1):87-92

[52] Giang DT, Pham V, Tanaka M, Teixeira J. Sterilization and germination of papaya (*Carica papaya* L.) seed and response to LEDs. Seed Science and Biotechnology. 2011;5(1):56-58

[53] Charuc JF. Evaluación de métodos de escarificación en semillas de Pacaína (*Chamaedorea* sp). Chimaltenango [thesis]. México: Universidad Rafael Landívar; 2016. 59 pp

[54] Fariñas J, Sanabria D, Silva R. Escarificación química de semillas de tres especies de *Centrosema* para sabanas bien drenadas. Zootecnia Tropical. 1997;**15**:221-237

[55] Chinea E, Barquín E, García A. Methods of seed germination in four leguminous forage shrubs. Sustainable grassland productivity. In: Proceedings of the 21st General Meeting of the European Grassland Federation. Badajoz, Spain; 3-6 April 2006. pp. 143-145

[56] Ávila E. Efecto de tratamientos, pregerminativos en la germinación de semillas de papaya (*Carica papaya*) [thesis]. Honduras: Zamorano; 2007

[57] Candiani G, Galetti M, Mendes V. Seed germination and removal of *Michelia champaca* L. (Magnoliaceae) in eucalypt stands: The influence of the aril. Revista Árvore. 2004;**28**:327-332. DOI: 10.1590/S0100-67622004000300002

[58] Norman, JC. Germination of *Michelia champaca* seeds after various short storage and seed coat treatments. Proceedings of the International Seed Testing Association. Vol. 36. 1971. pp. 335-342

[59] Bian F, Su J, Liu W, Li S. Dormancy release and germination of *Taxus yunnanensis* seeds during wet sand storage. Scientific Reports. 2018;**8**:3205. DOI: 10.1038/s41598-018-21469-9

[60] Contreras S, Barros M. Vigor tests on lettuce seeds and their correlation with emergence. Ciensia e Investigación Agraria (in English). 2005;**32**:3-10. DOI: 10.7764/rcia.v32i1.301

[61] Wang Y, Yu L, Nan Z. Use of seed vigour tests to predict field emergence of Lucerne (*Medicago sativa*). New Zealand Journal of Agricultural Research. 1996;**39**:255-262. DOI: 10.1080/00288233.1996.9513184

[62] Divsalar M, Oskouei B, Sheidaei S. Study on correlation of seed germination percent of two sweet corn hybrids with field emergence and some measured traits related to yield. Journal of Applied Environmental and Biological Sciences. 2016;**6**:44-50

[63] Rajjou L, Duval M, Gallardo K, Catusse J, Bally J, Job C, et al. Seed germination and vigor. Annual Review of Plant Biology. 2012;**63**:507-533. DOI: 10.1146/annurev-arplant-042811-105550

[64] Burton MG, Lauer MJ, McDonald MB. Calcium effects on soybean seed production, elemental concentration, and seed quality. Crop Science. 2000;**40**:476-482

[65] Hall R, Wiesner L. Relationship between seed vigor tests and field performance of Regar Meadow Bromegrass. Crop Science. 1990;**30**: 967-970. DOI: 10.2135/cropsci1990.0011183X003000050001x

[66] Maranville J, Clegg M. Influence of seed size and density on germination, seedling emergence and yield of grain Sorgun. Agronomy Journal. 1977;**69**:329-330. DOI: 10.2134/agronj1977.00021962006900020032x

[67] El-Keblawy A, Shabana H, Navarro N, Soliman S. Effect of maturation time on dormancy and germination of *Citrullus colocynthis* (*Cucurbitaceae*) seeds from the Arabian hyper-arid deserts. BMC Plant Biology. 2017;**17**:263. DOI: 10.1186/s12870-017-1209-x

[68] Erker B. Improve yield with high quality seed [Internet]. Fact Sheet No. 0.303. Crop Series. Production. Colorado State University Extension. 2014. Available from: https://extension.colostate.edu/docs/pubs/crops/00303.pdf [Accessed: 2019-05-05]

[69] das Graças Souza A, Smiderle O, Spinelli V, de Souza R, Bianchi V. Correlation of biometrical characteristics of fruit and seed with twinning and vigor of *Prunus persica* rootstocks. Journal of Seed Science. 2016;**38**(4):322-328. DOI: 10.1590/2317-1545v38n4164650

[70] Felizardo S, Freitas A, Marques N, Bezerra D. Características biométricas de frutos e sementes de *Oenocarpus bataua* Mart. com procedência de Almeirim, Pará. Revista Verde de Agroecologia e Desenvolvimento Sustentável. 2015;**10**:9-15. DOI: 10.18378/rvads.v10i5.3672

[71] Geritz S, Gyllenberg M, Toivonen J. Adaptive correlations between seed size and germination time. Journal of Mathematical Biology. 2018;**77**:1943-1968. DOI: 10.1007/s00285-018-1232-z

[72] Steinbrecher T, Leubner G. Tissue and cellular mechanics of seeds. Current Opinion in Genetics & Development. 2018;**51**:1-10. DOI: 10.1016/j.gde.2018.03.001

[73] Penfield S, MacGregor DR. Effects of environmental variation during seed production on seed dormancy and germination. Journal of Experimental Botany. 2017;**68**:819-825

[74] Teixeira JA, Rashid Z, Tan D, Sivakumar D, Gera A, Teixeira M, et al. Papaya (*Carica papaya* L.) biology and biotechnology. Tree and Forestry Science and Biotechnology. 2007;**1**(1):47-73

Chapter 2

Weed Seed Dormancy: The Ecophysiology and Survival Strategies

Jamal R. Qasem

Abstract

This chapter deals with seed dormancy of agricultural weeds, its definitions and types from the physiological and ecological point of view, and physiological and ecological factors inducing dormancy in different weed species. The role of different environmental factors, agricultural practices including herbicides application, selection pressure, and seasonal dormancy, weed density and population regulation, seed phenology, polymorphism, and modifications were emphasized. Factors induce or terminate dormancy and enhance seed germination and dormancy breaking have been mentioned and evaluated in addition to the ecological importance of seed dormancy and herbicide resistance, genetic bases of dormancy, and molecular studies were presented. The role of allelochemicals, stresses, and dormancy and their effects on seed longevity and germination regulation were thoroughly discussed. Dormancy breaking under laboratory conditions, role of plant hormones and other chemicals, and dormancy management in the field were reviewed in addition to information on seed dormancy/longevity and germination stimulants. Seed germination stimulants and inhibitors of parasitic weed and seed dormancy as a weed survival strategy were presented and discussed.

Keywords: dormancy, primary dormancy, secondary dormancy, weeds, ecophysiological factors, agricultural practices, dormancy-breaking chemicals, plant hormones, stress factors and dormancy, herbicide resistance and dormancy, dormancy management, stimulants and inhibitors, parasitic weeds

1. Introduction

Weeds represent a real persistent problem and can be found everywhere in all agricultural systems. They represent one of the main factors responsible for crop yield reductions, lower yield quantity and quality, and cause severe stresses and shortage in the supply of growth factors as they impair or negate crop yield. Losses caused by weeds exceed the combined losses resulting from insect and plant pathogens [1]. Weeds compete with crop plants for water, light, nutrients, and CO_2 under certain conditions. They harbor insect and plant pathogens and negatively affect water resources and the environment. Weeds have different life cycles and are grouped into annuals, biennials, and perennials. While perennials are mainly reproduced vegetatively, seed production is the main regenerative strategy involved in the succession of annual, biennial, and simple perennial weeds through the buildup and persistence of their seeds in the soil seed bank. Knowledge on weed seeds and their lifespan is essential for researchers as well as farmers in designing successful weed control programs. Seeds in the soil represent the passive weed population that remain viable for extended periods of time and able to re-infest agricultural lands in spite of effective weed control measures employed against both the active weed population found above the soil level and seed bank as well through certain control measures such as soil-applied herbicides, tillage, soil solarization, mulching, and flooding. Information on weed biology helps optimize weed management strategies by prediction of weed emergence time and weed infestation level and thus avoid unnecessary weed control input. Integration of knowledge on weed emergence and infestation level and seed dormancy status could be used to improve weed control strategies [2], while integrated approaches that place priority on depleting weed seed banks through interfering with dormancy or germination requirements have a strong potential to enhance weed management aspects of agricultural systems [3].

The main objectives of this study were to review the most recent advances on weed seed dormancy, highlighting the importance of weeds as the main agricultural problem, the importance of weed control and weed ecological and agricultural significance, and their importance in the agricultural system and in food production; emphasize the difficulties in weed control and challenges that the farmers face on what the weed species are possessing, role of seed dormancy in weed persistence and difficulties in weed control, role of genetic and ecological factors and their interactions, and influence of these on seed internal structure and physiology; and understand and introduce the readers to the recent findings on weed seed dormancy breaking and possible management under field conditions.

2. Importance of seed production and differences between weeds and crops

Weed seed bank is described as a reservoir at which both deposit and withdrawal operations occur. Seed production in terms of numbers is considered as a survival strategy that enables weeds to maintain their genetic lines and exist in the environment. It is important in agriculture since weeds can produce a huge number of individuals for ecological invasion and survival under unfavorable environmental conditions and thus maintain species where other regenerative propagules (e.g. vegetative organs for perennials) fail. Weeds are characterized by their huge number of seeds produced which is much higher than crop plants. These seeds are equipped with different modifications that enable their disperse far distances from mother plants to explore and invade new rich sites in growth factors and thus escape hazards in resource-depleted habitats underneath the parent plants. These modifications facilitate seed dispersal by different agents including water, wind, animals, machines, and packed agricultural materials and by man himself. However, the small-size or dustlike seeds of many noxious weed species do not require specialized agent for dispersal but can far-disperse by wind currents. In general, weed seeds are easily spread and transport from their origin, and some have found their way into the earth's planetary boundary. In order to maintain species genetic line, high seed production and seed modifications are necessary but not enough for species existence and persistence in a changing climate. Hence these characters must be accompanied with other mechanisms that help weed seeds remain viable and survive, and weeds grow and flourish in their habitats away from hazards including weed control measures. Therefore, seed production and modifications must conjugate with dormancy through which seeds of certain weed species such as *Lupinus* and *Chenopodium* can exist and remain viable for thousands of years. Dormancy is keeping seeds or buds safe until the cause of it is over. It is a significant feature contributing to weed survival rate and helps them avoid herbicides and other weed control measures along with unfavorable environmental conditions. The aforementioned weed characters are very well expressed and demonstrated when weeds also exhibit seed polymorphism, heteromorphy, or heteroblasty. Certain weed species produce different types of seeds per different plants or at different parts of the same plant, different in seed colors, structures, longevity, and more importantly germination capacity and requirements. Species of *Chenopodium*, *Amaranthus*, *Haloxylon*, *Xanthium*, *Rumex*, and many others are good examples (**Figure 1**). The ability of certain weed species to produce seeds of different colors, size, or coat characters in response to certain environmental conditions is very well documented. For instance, seeds of *Rumex vesicarius* L. are polymorphic (light and dark of various shades) and of a high potential viability [4]. Seeds are enclosed within showy, papery fruiting valves at maturation. Naked seeds exhibit non-deep physiological dormancy and usually require an after-ripening period for several months, after which they can germinate at any time of the year in a light-dark period. Light seeds are nondormant and show excellent germination in constant darkness; dark seeds are inhibited in darkness but not in a light-dark rhythm. The conditional dormancy of dark seeds is due to the pericarp that may restrict oxygen consumption by the embryo, contain chemical inhibitors, and/or impede radicle protrusion. A range of environmental variables is likely to affect the specific germination requirements of particular seed types. However, environmental conditions may induce secondary dormancy, in both light and dark seeds [4]. *Cirsium arvense* (L.) Scop. ecotypes showed differences in seed germination and in reproduction methods at different temperatures: one ecotype tends to reproduce vegetatively at high temperature (37°C), while the other is reproduced by seeds at low (17°C) temperature [5]. Pre-chilling releases seed dormancy of this species. Creeping thistle seeds did not show any apparent endogenous seasonal dormancy. Environmental conditions in the soil seed bank resulted from variations in germination. Dormancy was completely broken when imbibed seeds were stored for 2 months at 19°C, but at longer storage periods, dormancy was developed. Dormancy breaking at lower temperatures was slower and incomplete. Nitrate, desiccation, and light effects on seed germination were season-related.

In summary, prolific seed production, seed modifications, and dormancy are the characters jointly considered for a successful efficient weed species. These, however, are all absent in crop seeds hence exposed to different breeding programs resulting in loss of many characters that enable them to survive and tolerate harsh conditions

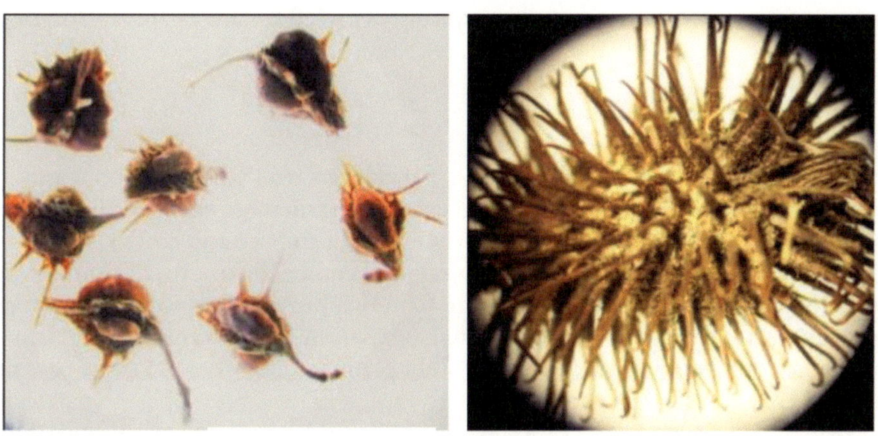

Figure 1.
Rumex acetosella L. (left) and Xanthium strumarium L. (right) fruits showing modifications on fruit case.

including plant disease, salinity, and drought resistance, and thus most crops lost their seed modifications and dormancy. Cultivated crops are well selected, and their seeds do not possess dormancy [6]. Therefore, the time of emergence and the number of established individuals could be simply determined by environmental factors (mainly temperature and moisture) [7, 8]. In contrast, prediction of weed seed germination and their emergence capacity are not possible because of dormancy. The number of established weed seedlings is strongly dependent on the dormancy level of the seed bank, and the emergence time depends largely on the seasonal dynamic variation in seed bank dormancy [9]. In addition to the mentioned characters, weeds continue producing seeds throughout their life cycle, set seeds at all growth stages, have seeds of different stages on the same plant, and show seed polymorphism.

3. Factors cause seed death in the soil

Upon the fall of mature seeds on the soil, these may be deeply deposited or remained on the soil surface and thus exposed to different climatic conditions including light and air temperature or to agricultural practices such as tillage, hoeing, or herbicides. Seeds on the soil surface may be inserted in the top soil layer 2–5 cm depth which is applied to seeds of most weed species and could be of great value facilitating rapid germination especially photoblastic species that require light for germination, or small seeds contain small food reserve. In addition, surface-laid seeds are liable to drift by wind currents or water erosion or disperse by different agents to new regions and thus avoid suffering from depleted resources under or in the vicinity of their parent plants giving them a survival value. Other soil-deposited seeds remain in full darkness and may be at different soil depths. These either germinate in full darkness or stay dormant if it requires light until brought to the soil surface by deep tillage. However, the ability of different species to emerge from different soil depths depends on their seed food reserve whether it is sufficient to support the seedling travel along the soil way distant or not. Some may consume all food storage before its emergence above the soil, and thus their growth is arrested during transit and dies. When conditions do not permit seed germination in the soil, seeds remain dormant, viable, and ready to germinate when these permit. The longevity of these seeds depends on the stored food and microbial attack of these in the soil. Other factors cause seed death including enzyme action and oxidation that denatures seed-stored food, protein coagulation, nuclei degeneration, and accumulation of toxic materials. In addition, seeds may be attacked by earthworms that collect weed seeds and move them into their burrows, while soil insects such as carabid beetles are voracious eaters and can consume a large quantity of weed seeds that drop into the soil.

4. Types of weed seed dormancy

Seed is simply defined as a fertilized egg produced after pollination, but apomixes, autogamy, and agamospermy also exist in certain weed species. The seed has been also defined as a ripened ovule consisting of an embryo and coats [10]. The embryo as the new plant in miniature is very well equipped structurally and physiologically for dispersal, has enough stored food that provide the growing seedling at early stages after emergence and until establishes itself for autotrophic organism or partially or completely dependent upon other plant species in case of some parasitic species (hemi- or holo-heterotrophic).

Dormancy in a general term is a state in which viable seeds, spores, or buds fail to germinate under favorable conditions of moisture, temperature, and oxygen for the seedling growth. It is referred to as an adaptive feature that optimizes the distribution of seed germination over time. It characterizes many weed seed populations; and this hampers efforts in predicting timing and extent of weed emergence. Indeed, the number of established plants of a weed is strongly related to the proportion of the seed bank that has been released from dormancy and the carrying capacity of the environment. Dormancy due to external conditions exerts influences on physiological and biochemical seed internal processes including enzyme activities, food transport to embryo, and metabolism or unknown internal factors or ecophysiological behavior that does not allow germination. Therefore, the causes of this status are due to the seed and its environment. On the other hand, germination involves the resumption of embryo growth and seedling emergence and growth. Germination requires moisture, oxygen, temperature, and maybe light in photoblastic seeds. Therefore, it proceeds whenever seeds are laid on a safe site to meet particular sets of environmental conditions which, presumably, are able to support not only germination itself but also to insure the survival and success of the offspring [11].

Several types of weed seed dormancy have been recognized and described under different terms as primary and secondary, inherent/genetic and environmental, innate, induced and enforced, constitutive and exogenous, and seasonal and opportunistic. Nikolaeva [12] mentioned 15 types of dormancy based on germination inhibitory and stimulatory factors. The primary dormancy (induced during seed maturation) and secondary dormancy (induced naturally or artificially following harvest) are mainly considered by most researchers.

However, based on the mechanism that causes dormancy, the following types are well recognized to occur in weeds:

1. Physiological mechanism of dormancy

2. Ecological or demographical consequences of dormancy

Both are important to understand the evolutionary adaptations that weed seeds have developed in the agricultural environment.

4.1 Physiological mechanism of dormancy

This dormancy includes the following three types.

4.1.1 Innate dormancy

Innate dormancy is also termed as a primary or genetic dormancy. It represents seed conditions when they leave parent plants in a viable state but not germinating although conditions are favorable mainly due to some property of the embryo or the associated endosperm or maternal structures. It is an inherited type that characterizes certain plant genera or families, and since genetically controlled, therefore its length depends on environmental factors. Seeds, however, will not germinate although conditions permit and dormancy period expires. The cause of such a dormancy includes a number of morphological and physiological factors, and these are as follows.

4.1.1.1 Hard seed coat

Hard seed coat that is impermeable or mechanically resists diffusion of water, oxygen, or both is also termed as physical dormancy [13]. It occurs in some or all species of the angiosperm families including Anacardiaceae, Bixaceae, Biebersteiniaceae, Cannaceae, Cistaceae, Convolvulaceae, Cucurbitaceae, Dipterocarpaceae, Geraniaceae, Lauraceae, Leguminosae, Malvaceae, Nelumbonaceae,

Rhamnaceae, Sapindaceae, Sarcolaenaceae, Sphaerosepalaceae, Surianaceae, and others; but it has not yet been reported in gymnosperms [14]. Leguminosae includes approximately 800 genera and 20,000 species that are widely distributed and adapted to different habitats [15, 16] and has a high frequency of physical dormancy [17]. Some examples of these are *Lupinus*, *Prosopis*, and *Vicia* that usually show water impermeability. Species of Amaranthaceae, Chenopodiaceae, Oleaceae, and Solanaceae do not possibly allow water diffusion to dissolve and transport food to the embryo for metabolic processes. The waxy cuticle is the major impermeable barrier to water entry (polyphenols and lignifications). It was found that seed coat impermeability to water is the major reason for the persistence of velvetleaf (*Abutilon theophrasti* Medik) seeds in the soil [18], while seed dormancy of *Avena fatua* L. could be easily broken by breaking the pericarp, scarification, or abrasion by sand. However, physical dormancy is caused by one or more palisade cell layer(s) called macrosclereids [19].

Other seeds do not diffuse oxygen for energy and embryo respiration and metabolic processes. Such seeds are said to be "gas-hard," for example, *Xanthium* species, at which seed testa prevents oxygen diffusion into the embryo and thus shows undormant long seed and short dormant seed. Impermeability to oxygen may be also due to the presence of mucilage in and around the seed coat and/or the consumption of oxygen by the seed coat itself.

In the third type of seeds, both water and oxygen are not diffused through the hard seed coat such as for *Convolvulus arvensis* L. seeds that can withstand 5 years soaking in water and species of the Chenopodiaceae family.

Hard seed coat may restrict the diffusion of O_2 to enter the seed, prevents the outward release of CO_2 and/or inhibitors from the seed or embryo and also embryo protrusion and expansion, and blocks light passage to the embryo. However, seed coat and other structures surrounding the embryo are extremely important for seed survival and germination regulation, since they protect the embryo against external hazards and regulate germination time.

In other cases, seed dormancy in many species is imposed by the structures surrounding the seed. In addition to the seed coat or testa, these also include the pericarp, glumes, palea (hull), and lemma in cereals. The palea, lemma, and pericarp are responsible for coat-imposed dormancy in *Avena fatua*. These may prevent water uptake and gaseous exchange, thus responsible on insufficient availability of oxygen to support the level of respiration needed for germination or to oxidize inhibitors. They contain chemical inhibitors in the seed cover that inhibit germination process or may prevent leaching of inhibitors from the seed. These structures also modify light reaching the embryo and blocking light penetration to the embryo and act mechanically to constrain embryo expansion. In *Datura stramonium* L. seeds, both the endosperm and to a lesser extent the testa impose dormancy on the embryos, hence restricting radical growth. Removal of the testa did not improve germination, but removal of the endosperm and testa enhanced germination, similar to soaking of seeds in gibberellic acid or benzyladenine solutions [20].

In order to enhance seed germination, seed coat must be destroyed mechanically or by microorganisms. However, in legumes the seeds are hard with thick-walled cells of testa surrounding the waxy layer.

In other cases seeds may fail to germinate because of mechanical resistance of the seed coat which can withstand a high pressure of 1000 Psi, such as for seeds of *Amaranthus retroflexus* L., *Brassica nigra* (L.) W.D.J. Koch, and *Capsella bursa-pastoris* (L.) Medik. *Clavaria major* C.F. Gaertn is another example at which hard pits require 623 kg to break.

In these species, the passage prevention or difficulty of water and oxygen inside the seed is not the cause of germination failure but may be enhanced to germinate by partial digestion of seed coat by animals and thus overcome their dormancy otherwise they extinct. Germination of many weed species was greatly improved after it passed through the digestive systems of animals and dropped out with animal feces mainly because their hard coat became lenient by secretions from the digestive system of these animals.

4.1.1.2 Presence of endogenous inhibitors

These are allelochemicals that prevent seed germination and cause self-inhibition (autopathy). Chemical inhibitors may be found on seed coat or its associated structures. *Chenopodium* spp., *Lactuca*, and *Beta* seed coats contain inhibitory chemicals. Inhibitors are also found in the embryo, cotyledons, endosperm, and some inside the seed. The outer coverings may prevent leaching of these inhibitors. However, when the embryo is isolated and placed in water, the inhibitor is leached out, and germination occurs as found in wild oat (*Avena fatua*) and *Xanthium* species.

The perianth associated with the seed coat of *Chenopodium murale* was the cause of its germination inhibition. Water leached from the perianth inhibited seed germination of wheat and barley [21], while water leached from the perianth-scarified seeds had no inhibitory effect on both crops. Inhibitors are also found in structures associated with the seed coat of *Chenopodium album* L. However, the period of effectiveness of these inhibitors depends on their stability. Removal of inhibitors may be achieved by placing seeds under running tap water or in irrigation channel in the field and thus washing off water-soluble inhibitors.

4.1.1.3 Control by biochemical trigger

In this case seeds need to be biologically stimulated. The photoperiodically operated triggers act through modification of the phytochrome system. Seeds of *Betula pubescens* Ehrh. require light and long-day photoperiod for successful germination. Dormancy breaking requires light and dark stimuli and also temperature stimuli. The effect of temperature on the rate of dormancy induction is not only dependent on prevailing temperature but also on temperature experienced by seeds during previous dormancy release and the resulting dormancy status of the seed population [22].

Chilling or temperature fluctuation may be also important. In two populations of *Papaver aculeatum* Thunb. exposed to temperature fluctuation, seed dormancy was weak, and fresh seeds germinated to nearly 100% at 20/10 and 25/15°C day/night if provided with light and up to 50% at 15/5°C, but germination was prevented at 30/20°C [23]. In another study, Dahlquist et al. [24] reported that six different weed species of different families showed different tolerance to lethal levels of high temperature and responded differently to germinate at different temperature regimes; however, temperatures of 50°C and above were lethal for seeds of all species.

Germination stimulants may be used under laboratory conditions but have little relevance to field situation. Gibberellins, thiourea, or nitrate ion in the soil solution could increase with soil temperature in the spring and thus could stimulate seed germination of *Chenopodium album* and *Avena fatua*.

4.1.1.4 Immature or rudimentary embryo when the seeds are shed

Embryo dormancy is defined as failure of a mature embryo to germinate or to grow even when isolated from the seed or dispersal unit and exposed to conditions favorable for growth. Embryo dormancy may be also exerted by cotyledons as for *Fraxinus* species seed that needs removal of the cotyledons or part of them to break the embryo dormancy. However, in this kind of dormancy, mature seeds of certain weed species are not able to germinate after leaving parent plants because the embryo is not fully developed. Seeds at this stage are physiologically immature when left by mother plants. For some weed species, the seeds require an after-ripening period for the embryo to get mature, and this requires in some cases changes in hormone contents or translocation of stored materials. Afterripening is the loss of dormant state over a certain period through exposure of seeds to a set of environmental conditions after maturation and separation from the parent plant [25]. The after-ripening process, however, occurs during a period of dry storage of freshly harvested mature seeds and is essential in releasing dormancy and in promoting germination [26, 27]. However, dry conditions are not always the case required, but plant species vary in environmental conditions that facilitate afterripening, while wild oat seeds after ripening under warm, dry conditions, and *Arabidopsis* and many other species respond best to cool, moist conditions [28, 29]. Seeds of *Acanthospermum hispidum* A.P. de Candolle that born immature and grow after the seed is shed and those of *Heracleum sphondylium* L. require several months after shed. Seeds of *Orobanche* spp. require a conditioning period of 3–4 months during which the embryo is mature enough to respond to host germination stimulants. This status, however, is overcome when embryo growth is complete.

Other weed species exhibit polymorphism and produce morphologically and physiologically different seeds that have different after-ripening periods such as for seeds of *Xanthium strumarium* L.

Several mechanisms sometimes may operate together in a single seed to break innate dormancy. For example, seeds of *Galium cracoviense* Ehrend. require lower Ca^{2+} ion concentration, moderate temperatures, and presence of light to germinate [30]. Garden cress (*Lepidium sativum* L.) seeds only germinate in response to a combination of light and temperature, while seed germination of *Chenopodium album* requires treatment with red light, cool temperature, and nitrate. The desert annual weed, *Trigonella arabica* Delile, has a dispersal unit equipped with at least four operating dormancy controls including water-soluble inhibitors, hard seed coat, sensitivity to light, and temperature.

Sometimes overcoming dormancy may be highly specific and adapted to certain conditions. Seed germination of *Rhus* spp. and *Epilobium angustifolium* L. and seedling growth occurred after forest fire that causes a waterproof layer of the dispersal unit to become permeable. Dormancy breaking may be also controlled by rain and temperature since there is an optimum levels of both to germinate.

Cold temperature may be required for germination of certain species; it could activate hormones and enzymes. Some seeds need exposure to alternating temperature between freezing degrees for several weeks to one or two exposures to high temperature. This temperature fluctuation causes heat shock and activates enzymes and hormones and enhances their mobilization and thus germination induction as the treatment for *Lithospermum arvense* L. seeds. However, many chilling requiring seeds do not show hard seed coat. Temperature can affect both germination and dormancy. At minimum and maximum temperatures, germination does not occur with some exceptions, while each weed species has an optimum temperature at which it germinates. Lower or higher than that germination is affected until totally prevented at extreme temperatures. *Amaranthus* sp. remains dormant at 20°C for 6 years, while raising the temperature released its dormancy breaking.

Light is another regulatory factor of seed germination for certain weed species having a light requirement (photoplasts) before the start of germination. Light-stimulated germination of seeds is known to involve the phytochrome system. The photoconversion of red light (P_r) to far-red light (P_{fr}) stimulates germination.

Lactuca sp. is a good example on light requirements for dormancy breaking. Seed germination occurs within a narrow temperature range, but giving light the seeds germinate promptly and uniformly over a wide range and under a variety of conditions that would inhibit germination in the dark. Dry seeds of *Lactuca* are insensitive to light, but when moistened and exposed to few foot candles for a few seconds, this treatment had a full effect on seeds. Moistened light-treated seeds retain their ability to germinate when dried and restored, but when subsequently remoistened, they germinate in darkness. However, the brief exposure to light that stimulates germination of *Lactuca* seeds is not enough for *Juncus maritimus* Lam. seeds to germinate. Continuous illumination works well with *Lactuca* but would inhibit seeds of *Atriplex rosea* L. plant that is fully stimulated by a brief exposure.

Sensitivity to the period of light and dark may determine the season of germination and growth, the flower initiation, and the end of bud dormancy. However, the value of light in stimulating or inhibiting seed germination may be very well demonstrated on species survival and existence knowing that germination inhibition by high light intensity may be of value in preventing germination and growth of winter weeds during the summer time at which soil surface may be exposed to unfavorable conditions such as drying or rapid seedling desiccation due to high temperature, high light intensity, and long photoperiod during summer and unsuitable for winter weeds. Conversely, germination inhibition of summer weed seeds during winter prevents their possible death by freezing temperature, strong cool wind currents, short photoperiod, and low light intensity which are not in favor of summer weed growth and survival. This kind of inhibition caused by light on the two weed groups of different growth requirements is a good example on the important value of light inhibitory effects for the survival of these species. However, seeds of many summer annuals at low temperatures under moist conditions provoke dormancy release, while high temperatures induce secondary dormancy. The seed dormancy level establishes the range of temperatures under which germination is possible [22].

4.1.2 Induced dormancy

It is an acquired condition of inability to germinate caused by some experience after ripening. This kind of dormancy is also called secondary dormancy as seeds are ready to germinate but may go into dormancy due to sudden changes in environmental conditions such as in temperature, moisture, and oxygen levels that cause physiological changes in seeds. Seeds, however, will not germinate, and dormancy exists even when conditions changed to favorable. This dormancy may be resulted from seed exposure to excessive light which lead to no germination in darkness, lack of moisture, high CO_2 pressure, low O_2 pressure, and deep seed burying that will not germinate until they are brought to the soil surface. However, other buried seeds by tillage may not germinate even after they are brought to the soil surface.

4.1.3 Enforced dormancy

This kind of dormancy is maintained in or on the soil or with seeds submerged in water. It is defined as the inability of seeds to germinate because of environmental factors. One or more factors necessary for germination are in a short supply or absent including the lack of moisture, low temperature, lack or low oxygen level, and poor aeration and unfavorable atmosphere. However, percentage of O_2 found in the soil depends on soil porosity, depth, presence of microbes, and amount of soil moisture. When the external limitation is removed as seeds are brought to the soil surface by tillage, they germinate. Sometimes this dormancy is due to placement of weed seed deeper than 5 cm in the soil by tillage. It results from the absence of red (r) light under the soil surface. Red light induces germination in seeds by activating their phytochrome system (P)-chromophore blue pigment attached to the protein molecule in the seeds. Far-red (Fr) light deactivates the system and thus induces dormancy in weeds. However, dormancy does not persist when the environment changes.

Both induced and enforced dormancy make the secondary dormancy. The importance of secondary dormancy became clearer as a survival strategy prevents seed germination when seeds are found deep in the soil and seedlings will not be able to emerge from deep soil layers. This kind of dormancy may be regulated through the phytochrome pigments found at low concentrations inside the seeds. These pigments when exposed to a high percentage of P_r/P_{fr} induce germination. The exposure time may be short enough for parts of the second dormancy. However, seeds from a single weed species may exhibit one or more types of dormancy or all three in succession over a period of time. Primary dormancy is found in the freshly shed seeds at which they will not germinate under any environmental conditions until dormancy is broken. After primary dormancy breaking, the seeds may germinate providing that conditions are favorable. If suitable external factors are not present, then secondary dormancy may develop. Secondary dormancy can be relieved and re-induced during many successive years [31] until conditions for germination become favorable. This phenomenon is called dormancy cycling [32]. However, physiological differences between secondary and primary dormancy are unclear [33].

From the ecological point of view, seed dormancy is also termed as a dispersal by time and is defined as an arrest in the development of seed embryo under external environmental conditions suitable for plant growth (phase more resistant to environmental hazards). It is critical for annuals not perennials. However, two approaches are prevalent, and these are as follows.

4.2 Ecological and teleological dormancy

Dormancy from an ecological perspective is defined as a seed characteristic that prevents germination, even if suitable germination conditions prevail, not involving embryo or seed morphology or germination mechanisms. This is either as follows:

4.2.1 Seasonal dormancy

This kind of dormancy occurred at which favorable factors for germination were found but seeds of certain plant species have winter or summer dormancy. It occurs in an environment where favorable growth conditions are seasonal and dormancy is usually clocked by solar rhythm. This is applied to all annual summer and winter weeds at which day length is important, while temperature may not be so if followed by cold weather. Day length is the best indicator of seasonal changes because it is a rather constant feature of the macro-environment. The disadvantage of seasonal dormancy is that seeds may not be developmentally advanced enough to take advantage of especially good spring or summer conditions. If the environment is not stable (rainfall in the desert, fire, soil disturbance), it may make conditions favorable for seedling growth, but the timing and duration of these events can be rather unpredictable.

Differences were found among populations of *Solanum nigrum* L. collected on two dates from different locations. Fresh seeds were conditionally dormant and germinated at higher alternating temperatures and in light, while seeds of *Solanum physalifolium* Rusby were deeply dormant. Seed dormancy is reduced during autumn, winter, and early spring in soil-buried seeds. The rate of dormancy release and induction is low at lower temperatures and increases as the temperature rises. High temperatures cause short-lasting breakage of dormancy followed by induction.

Seedling emergence of both species showed a bi- or three-modal pattern during an extended period in late spring and early summer. This enables the species to survive natural catastrophes or escape weed control operations. Dormancy is mainly induced during summer due to higher temperatures. This prevents seedlings from emerging too late and being killed by frost in autumn before reproduction [34]. Kołodziejek and Patykowski [35] reported that *Rumex confertus* Willd. Germination percentage and rate were significantly higher in light than in darkness. Seeds incubated for 12 weeks in the dark at 4°C exhibited secondary dormancy. Weed seeds undergo a seasonal deep dormancy in winter and early spring and a low level of dormancy in early autumn. Germination, however, decreased with soil salinity, while NO_3^- enhanced speed germination. Seeds burying at >0.5 cm reduced germination.

4.2.2 Opportunistic dormancy

In this kind of dormancy, seeds of certain species are able to take advantage from unpredictable environmental conditions or changes. It occurs when there is only a small seasonal element in the occurrence of favorable conditions; dormancy tends to be both imposed and released by the direct experience of the unfavorable or favorable conditions. For instance, deep tillage brings the seeds to the soil surface and thus would allow successful germination and establishment. Ephemerals in the desert sometimes take an advantage from the sudden rain shower during summer at which they germinate but later they suffer death because of the usual prevailing conditions of drought and high temperature in the desert during that period.

The advantage of seasonal dormancy is its predictable nature, while the advantage of opportunistic dormancy is its responsiveness. The differences between the two types are not exclusive but changed when conditions are changed. However, physiological description of dormancy may be a more valuable approach since the conditions of the embryo are what finally determine seed germination.

5. Physiology of dormancy in weed seeds

Dormancy is an adaptive trait that enables seed germination to coincide with favorable environmental conditions. From the physiology perspectives, gibberellins, ethylene, cytokinins, or abscisic acid (ABA) play an important role in inducing or inhibiting seed dormancy. The low level of ethylene is accumulated at the early stage of germination in seeds of different crops (e.g., *Ricinus communis*, *Lactuca sativa*, *Hordeum vulgare*) and weed species (e.g., *Avena fatua*) just prior to radical protrusion. Nondormant *Xanthium pensylvanicum* embryos produce more ethylene than dormant seed. Cytokinins increased in dormant seeds of *Rumex obtusifolius* and *Spergula arvensis* during dormancy breaking and chilling periods. On the other hand, ABA application inhibits seed germination. It has been suggested that related or common receptors for dormancy-breaking agents are present within the plasma membrane of the responsive embryonic cells. When triggered, these receptors initiate a signal transduction cascade, perhaps involving synthesis of or sensitization to germination-promoting gibberellins (GAs) that complete germination. Changes in the phosphorylating activity of membrane-associated, Ca^{2+}-dependent protein kinases that lead to dormancy or germination have been also proposed.

There is considerable circumstantial evidence that ABA is involved in regulating the induction of dormancy and in maintaining the dormant state. However, there is a paucity of unequivocal evidence that ABA is in fact an important controlling factor in the dormancy of most seeds. Dormancy is induced by abscisic acid during seed development on the mother plant. After seed shed, germination occurs due to reduction in the ABA level of the imbibed seeds because of ABA catabolism through 8-hydroxylation. ABA/gibberellins balance is the main environmental factor responsible for inducing or breaking seed dormancy. However, in different species, ethylene counteracts ABA inhibitory effects and stimulates germination. This effect is very well demonstrated in Brassicaceae seeds, which counteracts ABA effects on endosperm cap weakening, facilitating endosperm rupture and radical emergence. In contrast, ABA limits ethylene biosynthesis and action. Nitric oxide has been proposed to act against ABA inhibitory effects on ethylene and hence is produced rapidly after seed imbibitions and promotes germination by inducing the expression of the ABA 8-hydroxylasegene, *CYP707A2*, and stimulating ethylene production. The role of nitric oxide and other nitrogen-containing compounds, such as nitrate, in seed dormancy breakage and germination stimulation has been reported in several species. Both ethylene and nitric oxide have been shown to counteract ABA action in seeds, improving dormancy release and germination [36]. Abscisic acid has been also found to inhibit RNA synthesis. In seeds of *Chenopodium album*, ABA has been found to inhibit the embryo growth necessary to penetrate the coverings of the seed, although the initial events of embryo expansion are not prevented [37]. ABA deficiency was found to associate with the absence of primary dormancy, while high ABA content could promote seed dormancy. ABA is synthesized in the embryo and endosperm, and the balance between GA and ABA determines dormancy in weed seeds. GAs are known to obviate the requirement of seeds for various environmental cues, promote germination, and counteract the inhibitory effects of ABA, frequently in combination with cytokinins.

From the above information, it becomes clear that some plant hormones have roles in dormancy induction or breaking and thus inhibit or stimulate seed germination. Ethylene stimulates seed germination of several weeds such as in *Avena fatua*. Gibberellins increase in seeds, require stratification, and also facilitate degradation of food reserves in the endosperm or cotyledons necessary for germination.

Some chemical compounds or secondary metabolites are also known as allelochemicals such as phenolics, unsaturated lactones, short-chain fatty acids, coumarins, and many others have been reported as germination inhibitors present in seeds of many weed species [38, 39]. To enhance germination, leaching and oxidative destruction of these chemicals within the seed are necessary for dormancy termination. However, these allelochemicals may also play a positive role in seed viability and longevity since they prevent microbial attack and maybe destruction of weed seeds by soil pathogens and insects.

Ecological factors are involved in inducing dormancy or stimulation of seed germination and dormancy breaking. These factors include light, temperature, O_2, CO_2, and nitrate. Light causes weed seed dormancy. Some weed seeds require light in order to germinate, for example, *Galinsoga parviflora* Cav., *Portulaca oleracea* L., *Chenopodium album*, and *Amaranthus* spp. breaking dormancy is related to light that exists in promoting and inhibiting forms. Promoting form is favored by red light and inhibiting form by far-red light. Pigment generally initiates germination when in far-red light around 750 nm absorbing form and either prevents or has no effect on germination when in the form of red light around 660 nm. The inactive form (P_r) and red light, 650 nm, resulted in germination promotion, while the active form (P_{fr}) and far-red light, 750 nm, cause germination inhibition. Leaf canopy suppresses germination through shading effects since it promotes relatively low-red to far-red photon flux ratio, producing relatively low P_{fr} to P_r ratios in underlying seeds which inhibit seed germination. Great variations do exist between weed species in pattern of development of photoplastic properties.

However, light, moisture, temperature, and O_2 all act physiologically in enhancing or ending dormancy. Moisture or water is required to activate enzymes, compensate for water loss by the embryo through respiration, and dissolve and mobilize food into the embryo. Oxygen is necessary for aerobic respiration to provide energy for embryo growth, while water absorption, hormonal balances, metabolic processes, and germination induction will not proceed but only at certain suitable temperature. All factors, however, are required for biochemical and physiological activities that occur inside the seed including the living embryo.

6. ROS production and sensing in seeds

Reactive oxygen species (ROS) play an important role in seed life cycle. In orthodox seeds, ROS are produced at all stages in seeds active cells as well as in dry tissues during after-ripening and storage. ROS, however, are widely regarded as detrimental to seeds, but recent research results reconsider them as beneficial in seed germination and seedling growth. ROS regulate cellular growth, protect against pathogens, or control the cell redox status. They also act as a positive signal in seed dormancy release by interacting with plant hormones such as in transduction pathways of abscisic acid and gibberellins [40]. Different workers emphasized ROS roles in plant physiology and development under stress conditions mainly drought stress, and thus their production has been long considered as detrimental since it is linked with seed aging or seed desiccation, but they have also a positive important role in seed germination or dormancy release. They are important in metabolic activity during cell division, seed filling, seed survival at shedding, and seed rehydration and germination. ROS have an essential role in plant metabolism, energy production, and enzyme activities necessary to start seed germination and seedling growth. Their sensing and signaling role in seed different stages is evident. ROS is important in cell signaling in the dry state since it could accumulate during dry storage but would become actors of cell regulatory mechanisms only after seed imbibition. Oxygen is important in the guise of reactive oxygen species in further modulating dormancy and relaying environmental signals. Seed dry after-ripening is associated with the accumulation of ROS, resulting in targeted mRNA oxidation and protein carbonylation of transcripts and proteins associated with cell signaling (mRNA) and protein storage [41]. These modifications have been linked to dormancy changes during after-ripening and could underpin a mechanism indicating the passage of time. Recently the possibility of a further role for ROS to inform the seasonal response of the seeds through ultra-weak photon emission (UPE) has been suggested. It was hypothesized that beneath the soil surface the attenuation of light (virtual darkness: low background noise) enables seeds to exploit UPE for transducing key environmental variables in the soil (temperature, humidity, and oxygen) to inform them of seasonal and local temperature patterns.

7. Seed dormancy in response to stresses and herbicides

Seed germination is affected by many environmental factors, such as temperature, salt, light, soil moisture, oxygen concentration, and Ca^{2+} ions. Dormancy is a status to avoid and resist adverse conditions and must be evolved as a solution to the periodic, as well as nonperiodic, changes in the environment which impair the proper function of the plant during certain periods [42]. It may also prevent germination under apparently normal conditions, if they occur occasionally. In this way, it constitutes an evolutionary safeguard against the uncertainty of the environment. Drought, salinity, alternating temperature, photoperiod, burial depth, nitrates, nitrites and soil pH, artificial seed aging, agricultural practices, control methods, and radiant heat all influence weed seed dormancy.

Studies in controlled environments have already demonstrated that thermal conditions and, to some extent, water availability during seed set and maturation have an impact on the level of dormancy [43]. The level of dormancy in *Alopecurus myosuroides* Huds. seeds depends on the magnitude and timing of temperature and water availability during the reproductive growth phase. Water availability seems more important during maternal environmental perception and temperature during zygotic environmental perception [43]. Both temperature and soil moisture content are important factors in seed germination induction. Maximum (68–100%) and rapid (2.58 days) seed germination of *Calotropis procera* (Aiton, W.T. Aiton) occurred at 30°C but declined under water stress with increasing temperature, from 92.5 ± 1.1% at 20°C and 0 MPa to 2.8 ± 1.7% at 40°C and −0.4 MPa, respectively. Seeds were unable to germinate at ambient temperatures ≥40°C but remained quiescent and viable. Planting depth also influenced seedling emergence and water stress inducing a reduction in optimum germination temperature from 30 to 20°C. Short mean germination times increase seedling survival by rapid transition from endosperm resources to photosynthesis, whereas seed quiescence (cf. dormancy) optimizes germination opportunities in a semi-arid environment. Thus, the germination traits are likely promoted seedling survival and its spread [44].

Velvetleaf seeds germinated over a range of constant temperatures from 10 to 40°C regardless of light conditions, but no germination occurred at temperature below 5°C and beyond 50°C. Seeds germinated at alternating temperature regimes of 15/5–40/30°C, with maximum germination (>90%) at alternating temperatures of 40/30°C. Germination, however, was sensitive to water stress, and only 0.4% of the seeds germinated at the osmotic potential of −0.4 MPa. There was no germination at 0.6 MPa. Germination was also reduced by salinity and alkalinity stresses and did not occur at 150 mM NaCl or 200 mM $NaHCO_3$ concentrations. However, pH values from 5 to 9 had no effect on seed germination. The maximum seedling emergence (78.1–85.6%) occurred at 1–4 cm depth [45].

Bochenek et al. [46] reported differences between the cultivars of *Brassica napus* L. seed in their potential to exhibit secondary dormancy following environmental stress. A significant number of differences in gene expression between the cultivars were apparent in the transition from full-size embryo to mature seed. Most differences were apparent in the desiccation stage, and some were in genes related to signaling processes and protein biosynthesis. Authors suggested that the propensity of *Brassica* seeds to manifest secondary dormancy may be determined by changes in gene expression that occur during late seed development [47].

Breaking primary dormancy of achenes in *Cirsium arvense* only took place during the first stratification month at moderate temperature, which is mainly due to an increase in the average water stress tolerance in seed population. The induction of secondary seed dormancy during after-ripening at all temperature resulted mostly from a substantial loss of the seed's ability to tolerate water stress [46].

The effects of drought and herbivory on biomass and seed quality in *Vaccaria hispanica* (Mill.) Rauschert have been also studied by Cici [48]. The maternal water stress suppressed seed mass but stimulated seed dormancy in seeds. Progenies from the maternal stress environment were more persistent than those from the maternal control environment after being exposed to 45°C and 100% RH for 8 days.

Nassella trichotoma Hackel ex Arech. was identified to be a non-photoblastic, with germination percentages being similar under alternating light and dark and complete darkness conditions [49]. With an increase of osmotic potential and salinity, a significant decline in germination was observed. *N. trichotoma* seed dormancy break can be triggered by favorable alternating temperatures of approximately 25/15°C and ample water availability. Radiant heat has a positive effect on total germination. Osmotic stress and salinity significantly reduced germination, while water appeared as the most important limiting factor in germination. Soil pH is not a limiting factor on this species recruitment. Herbicide-resistant populations of the same weed species have been studied to identify differences in important environmental factors on its seed dormancy. The increase in osmotic potential and salinity caused a significant decline in germination. The pH had no effect on germination. Exposure to a radiant heat of 120°C for 9 min resulted in the lowest germination in the first population (33%) and in the second population (60%). In the burial depth treatment, both populations had the highest emergence of 1 cm depth. However, variation between the two populations was observed for the burial depth of 4 cm. Differences between populations were found in emergence and overall germination [49].

Seed germination of the salt-tolerant species, *Salicornia europaea* L., that produces dimorphic seeds of high salt tolerance limits has been studied by Orlovsky et al. [50]. Germination of large seeds was found to be 3–4 times higher than of small seeds under control and at 0.5–2% of all salts tested. Germination and plant growth in mixed sulfate-chloride salts were distinctly higher than in pure chloride salts; small seeds exhibited deep innate dormancy but stimulated by 0.5–2% of chloride and sulfate salts. Small seeds develop earlier, are more dormant, and are less salt-tolerant than large seeds. Seed dimorphism made the species more flexible in its response to varying salinity and more adapted to salt and temperature stresses.

GA_3 at concentration of 400 ppm strongly stimulated germination of *C. bursa-pastoris* at 12/12 h of light/dark and continuous darkness. While KNO_3 at 2 mmol had no effect on germination, long wet pre-chilling enhanced germination. Seed germination occurred at 10–30°C and within a range of pH of 3–11. On the other hand, drought and salt stress strongly inhibited germination, but authors suggested that weed seeds can germinate at high salinity. Sowing depth is critical for germination, and seedling emergence decreased with sowing depth. The rates of *C. bursa-pastoris* germination and seedling emergence were highest for seeds on the soil surface [38].

8. Dormancy and agricultural practices

8.1 Tillage

Tillage exposes seeds to light before reburial, allows greater diffusion of oxygen into and carbon dioxide out of the soil, buries residue, and promotes drying of the soil, thereby increasing the amplitude of temperature fluctuations and promoting nitrogen mineralization. These factors are known to terminate dormancy in several species. The effects of burial, however, on germination and longevity and of water stress and temperature on germination and dormancy induction of the weed *Sinapis arvensis* L. showed that soil-buried seeds exposed to high temperatures in summer broke dormancy [51], but low water potential and constant supraoptimal temperatures induced secondary dormancy. The threshold temperature for dormancy induction was about 19°C when water was available but decreased with reduced potential. Dormancy induction increased with burial depth and was induced to its highest level (96%) at a depth of ≥5.19 cm. Water stress or more burial depth can promote induction of seed secondary dormancy.

Tillage effects on seed dormancy of different weed species are very well demonstrated especially on photoblastic species. Tillage may affect P_{fr} and P_r ratios and germination induction or dormancy. This, however, is varied for different weed species. Chavarria [11] reported that under conventional tillage, *Amaranthus retroflexus* and *Digitaria sanguinalis* (L.) Scop. seeds tend to stay in primary dormancy or develop secondary dormancy, if they become buried; further soil disturbance promotes germination since seeds are exposed to light. Lower germination is expected for buried *Chenopodium album*, *Echinochloa crus-galli* (L.) Beauv., and *Setaria glauca* (L.) Beauv. seeds, but they may overcome this dormancy without further soil disturbance. Burying seeds of *Polygonum pensylvanicaum* L., in contrast, may result in an enhanced germination after experiencing low temperatures during winter, while non-geminating seeds of this species may enter into a secondary dormancy as induced by increasing temperatures [11]. Under no-tillage systems, seeds of all species except *Polygonum pensylvanicaum* may acquire an increased germination capacity in response to exposure to conditions such as light. For *Amaranthus retroflexus*, *Chenopodium album*, and *Echinochloa crus-galli* non-buried seeds, the germination patterns may be associated with a photoperiodic response. Accelerated after-ripening occurred in seeds of all species except *Polygonum pensylvanicaum*, when stored at increasing temperatures from 0 to 40°C. *Amaranthus retroflexus*, *Echinochloa crus-galli*, and *Setaria glauca* seeds germinated better in the dark, while *Chenopodium album* and *Digitaria sanguinalis* germinated better in the light. These differences, however, were less evident for seeds stored at temperatures above 20°C, which indicates an interaction between the temperature previously experienced by the seed and its response to light conditions during germination [11].

Weed emergence was also reported as increased following frequent, repeated tillage. Cultivation during daylight serves to increase weed populations. Daytime tillage increased seedling emergence of several winter annuals and doubled that in the night time tillage due to the extreme sensitivity to P_{fr} in buried seeds of certain weed species.

Tillage modifies soil temperature fluctuations or soil nitrate concentration, and continuous tillage depletes organic matter that leads to a change in soil color and thus modifies soil thermal regime. Tillage changes the position of seeds in the soil, while no-tillage leaves most seeds in the top 10 mm of the soil profile.

8.2 Fertilization and chemical applications

Nitrates affect seeds of several species and enhanced seed germination in the field. Nitrates may influence mother plant resulting in increased nitrate level in developing seeds. A strong correlation between nitrate concentration in the seeds and their germination capacity was also found. Nitrate and nitrite concentrations have been shown to stimulate dormancy release in some species although other species are released from dormancy by ammonium. Soluble N can stimulate germination of seeds of many weeds including *Amaranthus retroflexus* and *Chenopodium album*; manipulation of soil fertility has been extensively explored as a tool for reducing weed density [3]. Practices that avoid large pulses of soluble N early in crop development, such as delayed or split N applications, or use of slow-releasing N sources, such as mature compost, can delay weed emergence and reduce weed density in the crop.

8.3 Flooding

Under irrigation and flooding conditions, the soil has low oxygen concentrations. Low oxygen concentration terminates dormancy in seeds of some species including *Echinochloa turnerana* and *Leersia oryzoides* (L.) Sw. in rice crop. Flooding, however, causes death of unadapted species and thus facilitates the establishment of *Ambrosia tenuifolia* Spreng. because of the benefits from increased R/FR ratio.

8.4 Crop residue and burning

Thick layer of residue increasingly reduces and delays emergence, decreases temperature, and prevents light penetration [3]. Soil-incorporated crop residues yield allelopathic effects on weed seed germination. Decayed residues can immobilize large amount of N that consequently prevents termination of dormancy in some species. The stimulant effect of certain plant residues is also possible. Many plant-derived smoke components have been found to have a dormancy-breaking effect, and the role of nitric oxide has been identified.

Incorporation of legume cover crop materials and application of chicken manure can promote weed emergence and growth. For example, ammonium released from decomposing *Vicia villosa* Roth can stimulate germination of *Amaranthus hybridus* L. [52]. However, the effects of N fertilization on weed emergence are varied, owing in part to complex interactions between N and other factors such as ethylene concentration in soil, light, and genetic differences in responsiveness both between and within weed species [3, 53].

9. Seasonal dormancy and shift in population germination time

Seeds of certain weed species exhibit seasonal dormancy that allow them escape hazards that occur at a certain period in the year or unsuitable environmental conditions. This phenomenon is clearly demonstrated in annual, biennial, and perennial weeds. In annuals, summer-grown weeds will not germinate during winter since growth factors are not in favor of their germination, growth, and survival; the same is true for winter-grown weeds during the summer season. This is a danger avoidance strategy. In perennial weeds such as the woody spread *Prosopis farcta* (Banks and Sol.) Macbride, it inters into dormancy by the end of fall and winter seasons, drops all leaves, and stay dormant throughout the winter season. The weed breaks bud dormancy by the start of spring and grows vegetatively throughout the spring and summer seasons. The seeds of this weed also will not germinate during dormancy period [54]. The main factors control dormancy induction, or releases in all these species are the temperature and photoperiod.

Karssen [31] stated that seasonal periodicity in the field emergence of annuals is the combined result of seasonal periodicity in the field temperature and seasonal periodicity in the width of the temperature range suited for germination. Germination in the field is restricted to the period when field temperature (environmental factor) and the temperature range over which germination is possible (degree of dormancy) overlap. So, dormancy is related to the width of the temperature range over which germination can proceed and not to temperature in that range. Dormancy varies on a continuous scale, visualized by continuous changes in the range of environmental factors under which germination can take place [55].

Seeds of the winter annual *Bromus tectorum* L. lose primary dormancy in summer and are poised to germinate rapidly in autumn. If rainfall is inadequate, seeds remain ungerminated and may enter secondary dormancy under winter conditions [56]. Maximum secondary dormancy was achieved in the laboratory after 4 weeks at −1.0 MPa and 5°C. Seeds in the field became increasingly dormant through exposure to temperatures and water potentials in this range. They were released from dormancy through secondary afterripening the following summer. Different genotypes showed contrasting responses to dormancy induction cues in both laboratory and field. The changes in $\Psi b(50)$ can be used to characterize secondary dormancy induction and loss in this weed species.

A complex nature of responses may be exhibited by *Senecio vulgaris* L. in response to intense chemical selection. Seed germination and growth normally occur during the spring season (May–April) at which heavy application of simazine herbicide is practised. The repeated application of the herbicide during this growing period forced the weed to shift its germination time and life cycle to the winter annual where no herbicide application is practised. This shift in weed germination and growth is a strategy adopted by this species to avoid phytotoxicity of the herbicide [57].

10. Seed morphology, polymorphism, and dormancy

Seed polymorphism is an important factor in innate dormancy. It is a significant factor in spreading the germination of seed from the same plant over time and keeps farmers busy with weed control throughout the whole growing season. This phenomenon is widespread in the Amaranthaceae, Compositae, Chenopodiaceae, Cruciferae, and Gramineae families.

Genetic seed polymorphism is very well demonstrated in *Spergula arvensis* L. weed. The seed coat character is genetically controlled and associated with germination. Three different seed coat forms are found in weed seeds, each controls different levels of seed dormancy; these are the homozygous papillate form bearing about 120 papillae homozygous smooth coated form, and a heterozygous form which bears 60 papillae. The papillate seeds germinate readily at 21°C, while the reverse is true at lower temperatures [58]. This is an example of genetic polymorphism which is the production of seed of divergent morphology and behavior as a result of genetic segregation. Genetically controlled polymorphism distinctly showed different dormancy genotypes.

On the other hand, certain weed species show somatic polymorphism which is the production of seeds of different morphologies or behavior on different parts of the same plant. It is not a genetic segregation but a somatic one [58]. Among weeds showing such a phenomenon are *Xanthium* spp. Fruits of these species each contain a pair of seeds large and small one deeply dormant upper seed and a lower less dormant seed [59]. Dormancy breaking is different with the result of not less than 12 months of separate germination of the two seeds which could be regarded as an obvious insurance strategy. Dormancy breaking requirements are different for the two seeds.

Placement of *Datura stramonium* L. seeds on mother plants had a significant effect on seed germination. The middle and lower seeds on the maternal plant had less germination rate and seedling vigor than those in the upper part of the plant [60].

In *Avena fatua* and *Avena ludoviciana* Durieu, the grains are born on different parts of the individual spikelet that have different germination requirements. In *Amaranthus* spp. different seeds on the same plant are produced that are different in colors and seed coat morphology (**Figure 2**). *Chenopodium album* also produces brown to black seeds. Brown/black seed production ratio is 3–97%. Black seeds require cold temperature or supply of nitrate for dormancy breaking, while brown seeds are readily germinating at low temperature and are thin-walled. However, brown seeds germinate quickly, and seedlings are killed by winter cold or agricultural operations, but if they survive, they produce very large plants with a higher reproductive output, and then black seeds germinate in the spring season. Brown seeds ripened earlier giving the same ratio of brown to black seeds which is probably environmentally governed. Brown seeds represent highly opportunistic strategy, whereas black seeds are more seasonal and predictive in behavior.

In *Atriplex heterospermum* (Greene) Nels. & Macbr weed, black seeds are produced early in the season followed by large brown seeds to avoid unfavorable conditions. *Halogeton* sp. behaves similarly at which black seeds are produced in short days, while brown seeds in long days, and the weed shows differences in

Figure 2.
Amaranthus retroflexus black and brown seeds showing seed polymorphism.

dormancy-breaking requirements. Species of the Cruciferae family show seed size and color polymorphism such as for *Eruca sativa* Mill. and *Sinapis arvensis*.

11. Seed dormancy and herbicide resistance

Under conditions of herbicide application, some of these chemicals are absorbed by seeds or dormant buds, while others are not. These result in differences in germination, emergence, and growth patterns of different weed species. However, some herbicides may stimulate seed germination, while others inhibit this process or even kill the seed embryo. Differences also exist in hardness and permeability of the seed coat of different weed species at which species of Chenopodiaceae and Leguminosae are good examples on hard seed coat species. These characters cause differences in germination and growth of seedlings and may confer another cause of herbicide resistance. Avoidance of herbicide toxicity may result from seed interring into dormancy and not further responding to the applied herbicide with no absorption or translocation of the herbicide into the embryo. In addition, herbicide molecules may be deactivated or degraded inside the seed itself by some oxidative enzymes or may bound into certain constituent inside the seed. On the other hand, stimulation of weed seeds to germinate using certain herbicides also exists and allows higher seedling emergence and partitioning of herbicide molecules among individuals of weed species. Division of herbicide molecules among the high number of emerged seedlings would further dilute herbicide inside weed plants. All the above mentioned factors should be considered when herbicide resistance is discussed. These may cause great differences in weed seed germination, seedling growth patterns, and distribution in the field. Seed germination rate was often more rapid for herbicide-resistant *Echinochloa oryzicola* Vasing. than for herbicide-susceptible seeds, implying greater dormancy in the latter. Germination rate was often more rapid for herbicide-resistant than for herbicide-susceptible seeds, implying greater dormancy in the latter [61]. Population shift in life cycle has been previously discussed as shown by *Senecio vulgaris* L. in order to avoid herbicide phytotoxicity. Scursoni et al. [62] have shown that seeds from *Avena fatua* plants that had survived the application of diclofop-methyl in barley crops had a lower dormancy level. Moreover, an anticipated emergence timing of *A. fatua* was detected in plots that had been treated with diclofop-methyl the previous year in relation to the emergence timing observed in plots that had not been treated with the herbicide.

12. Factors enhance seed germination and dormancy breaking

Dormancy is synchronized to the environment by a complex regulatory system. This is directed by the balance between the dormancy-promoting (abscisic acid) and dormancy-releasing (gibberellins) hormones via both hormone levels and sensitivity. Seed dormancy is a survival mechanism underlying the life cycle strategies of plants by controlling the seasonal timing of germination in the natural environment.

Weed species differ widely in seed dormancy and longevity, season in which they emerge and grow, depth from which they can emerge, and seed responsiveness to light and other germination stimuli. Weed seeds that remain dormant in the soil often germinate in response to changes in temperature, moisture, oxygen, or light.

Overcoming seed dormancy may be easily achieved under laboratory conditions through a number of practices/treatments including seed hand scarification, rubbing and peel/cortex/extra structure removal, alternate temperature, chemical treatments, and dipping/soaking seed in water (legumes) or could be changed by scarification—with acids or microbes, rubbing on a sandpaper or pricking it with a pin or needle. As an example, *Rhynchosia capitata* (Heyne ex Roth) DC. seeds exhibit physical dormancy that is mainly due to the impermeability of their coat. Mechanical scarification and acid scarification (soaking of seeds in H_2SO_4 for 60 and 80 min and in HCl for 12 and 15 h) were very effective in breaking dormancy and enhancing germination [63]. Dormancy can be also overcome by alternate treatments of wetting and drying aiming to destroy the hard seed coat. In addition, seeds may be frozen and then dissolved, exposed to cold temperature and stratification, or washed by water for removal of inhibitors. Seed germination of *Eclipta prostrata* (L.) L. was completely inhibited in the dark, but in the dark/light, it was 93% at 20/30°C alternating day/night temperature. Germination was more than 80% at 140°C followed by incubation at 30/20°C for 14 days but declined at higher ratio until zero germination at 200°C. Germination is tolerant to salinity but highly sensitive to water stress. Seeds germinate at pH of 4–10 and are enhanced closure to soil surface but reduced thereafter until no germination obtained from a depth of 0.5 cm [64].

A novel method that overcomes coat-imposed dormancy has been reported by Tiryaki and Topu [65]. In their method, seeds were stored at −80°C for certain period and then treated immediately with hot water of 90°C for 5 s. This approach of freeze-thaw scarification provided 84 and 75% germination compared with 3 and 26% of *Lupinus albus* L. and *Trifolium pratense* L., respectively. In other cases genetic-controlled dormancy may not possibly overcome by physical or mechanical treatments as for seeds of *Nicandra physaloides* (L.) Gaertn. which possess one pair of iso-chromosomes. The presence or absence of an iso-chromosome determines whether a seed will germinate readily (2n = 20) or is dormant (2n = 19). *Echinochloa oryzicola* dormancy is possibly relieved through stratification. Stratification temperatures, moisture levels, and durations contributed to its seed dormancy released by decreasing hydrotime required for germination and by eliminating any germination sensitivity to oxygen [61]. Alternating temperatures nearly doubled germination rate in all weed populations of this species. Stratification at Y = 0 MPa increased germination rate compared to stratification at lower water potentials. Maximum germination rate occurred after 2–4 weeks of stratification at 0 MPa; it was suggested that field soil saturation in winter would contribute toward *E. oryzicola* dormancy release and decrease the time to seedling emergence.

Phleum paniculatum Huds. seeds had a shallow dormancy (20–30 d) when stored at room temperature (25 ± 5°C). Seeds could germinate at constant temperatures between 10 and 25°C. Light was not essential for seed germination, and pH values from 4 to 10 did not inhibit germination. Seeds were moderately adaptable to water potential and NaCl concentration, and germination rates decreased by 50% when water potential was −0.4 MPa or NaCl concentration was 130 mM. Increased soil burial depth decreased the seedling emergence, and no seeds emerged at depth more than 4 cm [66].

Cold stratification may be a highly selective treatment and hence did not decrease dormancy for any of *Lamium* species collected in Sweden, while warm stratification (21°C) of seed batches of all species that germinated when fresh increased dormancy but decreased over time for all tested species. Pretreatment with dry storage was most efficient in reducing dormancy for all annual *Lamium*. In contrast to cold or warm stratification, dry storage led to germination in darkness for all species [67]. Warm stratification increased germination, both in darkness and at 15/5°C, but did not cause germination of any of the two *Papaver aculeatum* populations tested at 30/20°C. Cold stratification reduced germination and limited germination to the cooler temperatures. Alternating cold and warm stratifications showed that the species undergoes dormancy cycles [23].

Seed dormancy of *Polygonum persicaria* L., *Chenopodium album*, *Spergula arvensis*, and *Sisymbrium officinale* (L.) Scop. [68] is regulated by field temperature and not nitrate and light. *Aciphylla glacialis* F. Muell. ex Benth seeds possess physiological dormancy that is overcome by cold stratification. Seeds have undeveloped embryos at dispersal but grow to germinate after 4–9 weeks at both constants 5°C and 10–5°C (day-night) temperatures. The species exhibits morphophysiological dormancy at which final percentage germination and dormancy status varied significantly among natural populations [69]. In another study Rhie et al. [70] reported that warm stratification (25/15°C) stimulated embryo growth and germinated the seeds of bamboo (*Nandina domestica* Thunb.), but cold stratification (5°C) was not required. Warm stratification at a constant 20°C speeds seed germination more than a fluctuating temperature at 25/15°C and shortened germination time by 4 months compared with that under natural conditions. Gibberellic acid (GA_3) at 100 and 1000 mg L^{-1} could substitute for warm stratification and break dormancy in seeds incubated at 15/6°C. In addition seeds may be exposed to light for chemical changes inside the seed or can be treated chemically with germination stimulants such as potassium nitrate, gibberellic acid, cytokinins, and auxins.

In the field, many agronomic practices affect weed seed dormancy and germination through their effects on the microenvironmental and edaphic conditions surrounding the seeds in the soil. Light penetration, soil water content, soil fertility, and temperature are modified by tillage, planting, harvesting, and other production practices that enhance or prevent weed seed germination. Changes in these environmental factors may modify indirectly phytohormone concentrations during seed development, which can subsequently affect dormancy status of the mature seed [3, 71].

It has been postulated that temperature is the only factor that directly influences the dormancy state of seeds, although the effects of other factors such as nitrate and light cannot be excluded [72]. Germination is influenced by factors such as temperature, light, nitrate, gaseous environment of the seed, and moisture content. Light appears as the most suited factor to influence in the field to enhance weed control. The behavior of weed seeds, in terms of dormancy characteristics, can be substantially different according to the location of the seed in the soil. Much of this response is related to a phytochrome requirement for germination, specifically, to the interconversions between the active (P_{fr}) and inactive (P_r) forms while the physiological mechanism involved in this conversion is poorly understood.

The changes in dormancy and germination of *Chenopodium album* L., *Polygonum persicaria* H., *P. lapathifolium* L. subsp. *lapathifolium*, *Sisymbrium officinale* (L.) Scop, and *Spergula arvensis* L. found regulated by temperature. Soil moisture and nitrate content had no effect. Germination of *C. album*, *S. officinale*, and *S. arvensis* was stimulated by light, nitrate, and desiccation. These factors all increased the width of the range of temperatures over which germination could proceed and therefore affected the expression of dormancy. Germination depended on the one hand on the actual field temperature after exhumation and on the range of the germination temperature, which was determined by the dormancy status of the seeds and soil temperature during burial. When nitrate was added during the test, the germination temperature range became wider, and germination could occur during a longer period of the year [73].

Karimmojeni et al. [74] studied dormancy breaking in seeds of *Thlaspi arvense* L., *Descurainia sophia* (L.) Webb ex Prantl., and *Malcolmia africana* (L.) W.T. Aiton (Brassicaceae) and reported that burying seeds of *D. sophia*, at a depth of 10 cm for 60 days, resulted in 55% germination and seeds dry-stored at 20°C for 180 days (45%) showed the highest level of germination. In *M. africana*, the germination percentage reached 95% when seeds buried at a depth of 1 cm were soaked in a GA_3 concentration of 150 ppm. *T. arvense* had the lowest level of germination compared to the other species. The highest percentage of *T. arvense* germination was obtained in seeds treated with 150 ppm GA_3. Potassium nitrate partly increased germination of *M. africana*, which initially was less dormant than those of *T. arvense* and *D. sophia*. Light is the most common germination trigger, though many seeds also respond to temperature and moisture fluctuations, increased aeration, and increased release of nitrate and other soluble nutrients that occur in tilled soil [75]. Tillage may end dormancy stage by exposing seeds to temperature and light. In many weed species, dormancy may be ended by treatment with a single factor or a mixture of ecological factors such as exposure to cold or humid conditions or to cold temperature and light.

In certain species cracking of the hard seed coat resulted from its exposure to mechanical pressure through coldness, freezing, scarification, or abrasion or through microbial attack that is necessary for germination. In other species seed coat must be destroyed, modified, or partially dissolved to provide the embryo with the necessary secondary growth factors. Embryos that are constrained by a mechanical barrier, such as the surrounding endosperm, perisperm, or megagametophyte (such as those that exhibit coat-enhanced dormancy), appear to require a weakening of these structures to permit radicle protrusion. This weakening involves partial enzymatic degradation of the cell walls. Seeds of *Datura ferox* exhibit increased endo-p-mannanase and p-mannosidase activities in the micropylar region of the endosperm after red light stimulation and many hours before the radicle protrudes through it [76]. Sometimes it is necessary washing the inhibitory chemicals. In other cases humid cold period or stratification is required during autumn and late in winter.

Certain species such as wild oats will not germinate unless seeds are exposed to warm dry conditions for dormancy breaking. Many weed species require light for seed germination, and this occurs when they get mature or may be stimulated by seed burying in the soil. Therefore, one effect of tillage is through dormant seed exposure for red light even for a short time to enhance germination. Some researchers study the implementation of tillage during the night to prevent stimulation of seed germination of certain weed species. The ratio of the active phytochrome far-red (P_{fr}) and inactive phytochrome red (P_r) determines whether the seed is dormant or not. The optimum ratio is established after exposure of the seed to white light which converts P_r to P_{fr} in the embryo. Outer coverings form a filter to high incident light on the seed and modify its effectiveness in converting P_r to P_{fr}. *Chenopodium album* seeds with dark seed coats are less responsive to light than their thin, light-coated counterparts. Dark seeds filter out light and reduce the conversion of P_r to P_{fr} when exposed to light; hence they remain dormant for longer periods than thin light-coated seeds.

In general, the main factors inducing or ending seed dormancy are light including day length, light type, dark period, and photoperiod; immature embryo; impermeable seed coat to water, oxygen, or both as in seeds of *Amaranthus* spp., *Capsella bursa-pastoris*, and wild *Brassica*.; and chemical inhibitors such as ABA and allelochemicals. Oxygen of percentages in the soil ranged between 1 and 9%. CO_2 with a percentage between 5 and 15%, temperature. Low oxygen prevents oxidative stress and germination, while high CO_2 induced dormancy of *Brassica alba* seeds. The after-ripening needs which are irrelevant to the degree of the embryo maturation are also involved in dormancy due to physiological changes that are not very well understood.

Soil disturbance and light stimulate germination and emergence prior to crop planting in order to remove as many weeds from the soil seed bank as possible. However, weed seed banks can be manipulated [75] by encouraging seed germination or trapping them to germinate, modifying their environment, placing seeds in a position that are not able from or in others where they are much exposed to heat and light, and using certified crop seeds.

A dense, shading plant canopy can also deepen the dormancy in some weed seeds. The dim green light under such a canopy can actually be more effective than continuous darkness in inhibiting the germination of light-responsive seeds [77]. Light quality under such foliage may have rendered weed seeds more dormant [78]. Dense crop canopies may also reduce subsequent weed emergence by reducing seed production or increasing seed mortality and hence provide favorable habitat for seed predators, resulting from reductions in the seed bank and subsequent weed emergence [79]. This dormancy strategy works best for annual weeds whose seeds often show conditional, light-mediated dormancy. Drought and shade were found to reduce reproductive allocation and resulted in seed of *Avena fatua* with less intense primary dormancy than the plants grown under resource-rich conditions, but had no apparent effect on seed vigor [80].

Karrikinolide may be an efficient means of stimulating weed seeds to germinate. These weed seeds would otherwise remain viable in the weed seed bank. Karrikinolide appeared to stimulate a broad spectrum of weed species, including wild turnip, wild radish, wild mustard, wild oat, cape weed, barley grass, and Paterson's curse. Germination stimulation of weed seeds was dependent on seed dormancy state, with some species (i.e., wild turnip and barley grass) responding differently depending on seed maturity conditions in the maternal environment. The application of karrikinolide at relatively low rates (2 g/ha) to weeds both *ex situ* and *in situ* can significantly improve weed seed germination while lower rates were inefficient. Karrikinolide seems working as a germination stimulant rather than a dormancy-breaking agent [81]. The smoke-derived chemical karrikinolide responses of seeds differed among populations of *Brassica tournefortii*, but this variation was reduced when seeds developed in a common environment. The KAR1 responses of each population changed when seeds developed in different environments. Different parental environments affected germination responses of the populations differently, showing that parental environment interacts with genetics to determine KAR1 responses. Seeds from droughted plants were 5% more responsive to KAR1 and 5% less dormant than seeds from well-watered plants, but KAR1 responses and dormancy state were not intrinsically; however, the parental environment in which seeds develop is one of the key drivers of the KAR1 responses of seeds [82].

13. Parasitic weeds and seed germination stimulants and inhibitors

13.1 Germination stimulants

Natural chemicals may include seed germination and growth stimulants including those for parasitic weeds. The ability of these chemicals to modify or break down seed dormancy and physiologically activate food transport, and embryo growth and development, or ruling out the over wintering stage of different living organisms, defense mechanisms, parasitic plants attachment and historian invasion to host tissues are examples on their positive effects [83].

Several natural chemicals have been identified as seed germination stimulants of parasitic species [84]. The main groups are the sesquiterpene lactones [85, 86]; some are alectrol from *Vigna sinensis* L., orobanchol from *Trifolium pratense* [87], and strigolactones and orobanchol from sorghum [88, 89]. Strigol was identified from the root exudates of *Gossypium hirsutum* L. plants and *Zea mays* L., which all are also produced by *Striga* host plants [90]. Some sesquiterpene lactones induce seed germination of *Orobanche cumana* Wallr. [91] better than the synthetic germination stimulant, GR24. However, Rice [39] stated that the stimulatory effect of allelochemicals is rather very limited or even rare and usually associated with low concentration effects of the compounds.

Strigol, alectrol, and sorgolactone are *Striga* germination stimulants, all produced by the parasite host plants, while different plant species have shown a strong ability to stimulate seed germination of different *Orobanche* species by more than 90% [90]. Ethylene was reported by Zehar and Fer [92] to induce germination of GR24-conditioned seeds of *Orobanche ramosa*, and its synthesis is required for parasite germination. However, a large number of synthetic substances were reported to stimulate germination of *Striga* species, among which are coumarin derivatives, scopoletin, thiourea, allylthiourea, sulfuric acid, sodium hypochlorite, and inositol [93]. Thidiazuron herbicide has been reported to activate ethylene release [94] and thus indirectly enhance *Striga* seed germination, although it is regarded by the same author as an inhibitor of haustorial development.

Ethylene- and ethephon-releasing compounds were effective in stimulating *Striga asiatica* (L.) Kuntze [95], while under field conditions, it was highly efficient against *Striga hermonthica* (Del.) Benth. Other chemicals have been found to stimulate *Striga* germination *in vitro* including several growth hormones and kinetins. Thuring et al. [96] were able to synthesize two diastereomers of demethylsorgolactone. Nijmegen-1 was active at low concentration as a suicidal germination agent for both *Striga* and *Orobanche* [97]. Yoneyama et al. [98] reported that jasmonates and related compounds elicited seed germination of *O. minor* and methyl jasmonate was the most active stimulant on several *Striga* and *Orobanche* species. The synthetic germination stimulants, strigol analogues (GR compounds), stimulate seed germination of different *Striga* and *Orobanche* species. Strigol has been reported for *Orobanche* stimulation [99], and its analogues GR7 and GR45 stimulated germination of *O. minor*.

Luque et al. [100] reported that parthenolide and 3,5-dihydroxydehydrocostus lactone significantly increased *O. cumana* germination and exhibited higher activity than GR24. The positive role of this hormone was further confirmed. The activity of germination stimulants for suicidal germination of *Orobanche* seeds under field conditions could be substantially enhanced by applying brassinolide (2a, 3a, 22R, and 23R-tetrahydroxy-24Smethyl-B-homo-7-oxa-5-a-cholesten-6-ono) and related compounds to the infested soils [101].

The role of microorganisms (e.g., *Streptomyces*) has also been implicated in *Orobanche* seed germination. Christeva and Naumova [102] reported different strains of *Streptomyces* to induce germination of *O. ramosa* and concluded that microorganisms may take part in germination processes of this parasite. In another study, Yoneyama et al. [103] reported stimulation of *O. minor* seed germination by certain fungal metabolites including cotylenins and fusicoccins at concentrations as low as 10^{-5} M and up to more than 50%.

Different plant species have been reported to show a strong ability to stimulate seed germination of different *Orobanche* species by more than 90%, and extracts of hundreds of plant species were tested for possibly stimulating or inhibiting seed germination of different *Orobanche* spp.; many proved effective and may be considered as trap, cover, and catch species or a source of natural germination stimulants for these parasites [104–106].

13.2 Germination inhibitors

Inhibitory allelochemicals may work at different stages of the parasite's life cycle from germination to growth and development. However, a reverse action may be obtained at low concentration received by the parasite.

Serghini et al. [107] reported that coumarins affect the normal growth and development of *Orobanche cumana* seedlings and the effect was more intense from the resistant than in the susceptible cultivars. Coumarins (ayapin and scopoletin) from sunflower plants were implicated as inhibitory allelochemicals to germination; growth and development of O. *cumana* [91] are both excreted by sunflower roots into the environment and could act as toxic allelochemicals to the parasite [108]. Bar-Nun and Mayer [109] reported the presence of large amounts of oleic and linoleic acids and small amounts of stearic and palmitic acids in *Orobanche aegyptiaca* (Pers.) Pomel. seeds. The authors expected these acids to interfere with sugar metabolism and thus prevent parasite seed germination during the preconditioning period. Sunflower seeds treated with 40 ppm of benzo(1,2,3)thiadiazole-7-carbothioic acid S-methyl ester (BTH) for 36 h completely prevented infection of *O. cumana* in root chambers [110]. In pot studies, at considerable inoculums of O. *cumana* seeds, the total number of parasite shoots was reduced by almost 90% with 60 ppm of BTH. Different *Fusarium* spp. were reported by Sauerborn [111] to infect *Orobanche* and *Striga* seeds and parasite plants, and he emphasized the role of phytotoxins in the process. He added that members of the genus *Fusarium* produce toxins such as enniatin, fumonisin, fusaric acid, moniliformin, and trichothecenes that possess a broad range of biological activities and metabolic effects. However, some of these compounds have been already proposed and considered to be used as natural herbicides [112]. Later studies showed fusaric acid and 9,10-dehydrofusaric acid to be active at low doses in reducing *Striga* seed germination [113]. Zehar and Fer [92] found T-2 toxin from *Fusarium* sp. to be the most active among 14 fungal toxins tested, inhibiting 100% seed germination at 10^{-5} M and was active down to 10^{-7} M (19% inhibition). Deoxynivalenol [vomitoxin], produced by *Fusarium* spp., was also very effective, causing 100 and 69% reduction in germination when assayed at 10^{-4} and 10^{-5} M, respectively. The high activity shown by some fungal toxins suggests that they may have potential as more natural and safe herbicides to suppress *S. hermonthica* seed germination. The use of toxic secondary metabolites could represent a useful alternative strategy in the management of parasitic weeds, by interfering with the induced germination process, and that fungal culture extracts could be an interesting source of new compounds acting as natural and original herbicides on these parasites.

Ancymidol, uniconazole, and paclobutrazol were reported as strong inhibitors of *Orobanche ramosa* germination, and CCC, daminozide, and prohexadione at 10^{-5} had a similar effect [92]. The inhibitory effect of paclobutrazol and uniconazole could be resulted from the increased level of abscisic acid.

Oxidative metabolism of ABA into phaseic acid and exogenous ABA is a strong inhibitor of *O. ramosa* germination [89]. Germination of *S. asiatica* was reduced by silver thiosulphate and COCH, inhibitors of ethylene action, and ACC oxide [114]. Aminoethoxyvinyl glycine reduced cytokinin thidiazuron-induced parasite germination.

14. Genetic studies on weed seed dormancy

Seed dormancy is mainly found in wild species in which weeds form the integral part of these species and are facing extreme challenges under field conditions. Dormancy is a strategy of weed survival and persistence that challenge farmers under all conditions. In contrast, crops lack such a trait and always show rapid and uniform seed germination [6] with some exceptions as for cereals that possess a moderate degree of dormancy to resist preharvest sprouting (the germination of seeds after maturation but before harvest in moist environment) that results in substantial yield loss. Dormancy is a genetically complex trait controlled by polygenes, but its effects are influenced by the genetic background and environmental factors [115]. However, genotype-by-environment interactions have been reported for seed dormancy in different species [116, 117]. The growth environment greatly affects both the number and the influence of individual quantitative trait locus (QTL) in a mapping population [118]. Gu et al. [119] suggested the presence of genetically complex networks in the regulation of variation for seed dormancy in natural populations of weedy rice (*Oryza sativa*). Multiple loci and epistasis control genetic variation for seed dormancy in the weed. Iso-chromosomes have been also mentioned to determine seed germination and dormancy. However, molecular studies on dormancy genetics are clearly rare, and there is a need for research in this aspect and genetic dormancy differences among and between weed species and their populations and the link of these with environmental conditions.

15. Seed dormancy as a weed survival strategy

Dormancy is a property that enables weed seeds to survive conditions hazardous to plant growth, such as the periods of extreme heat and drought in certain geographical regions or the long cold winters in temperate regions, and allows them to germinate at some later time or in some other place [120, 121]. Similarly, Roberts [122] indicated that seed dormancy mechanisms tend to inhibit seed germinating at the wrong time and in the wrong place. Hence, weed seeds can persist in the soil for many years and germinate after experiencing conditions favorable for seedling survival through maturity [120]. Such a behavior results in the accumulation of large quantities of seeds in the soil, forming either transient or persistent banks which constitute the regenerative strategy developed by many weed species [31, 120]. In further support of this concept, the analysis of the composition of most seed pools has revealed that dormant seeds are only produced in large numbers by species whose growing populations are subject to periodic local extinction, such as in the case of early succession. In addition, species exposed to elimination conditions such as the use of extensive applications of herbicides or implementation of certain other agricultural practices (flooding, soil solarization, deep tillage, and continued disturbance) tend to show tolerance or resistance to such practices and deep seed dormancy as a hazard avoidance strategy. Similar strategy is also expressed well under severe stress conditions such as extreme drought and salinity.

16. Conclusions

Seed dormancy is of different types and has several definitions; it is a highly complicated phenomenon weakly understood until now in spite of the huge number of publications available. The poor understanding of dormancy is mainly due the complexity of the factors involved including mechanical, physiological, and biochemical that may be also genetically and/or environmentally controlled. Although much is known on dormancy induction and breaking, but the complicated and interrelated issues occur in the seed itself including the seed coat, embryo, cotyledon, endosperm, cell organelles, nuclei, and associated structures that all need much research work on their roles and effects on dormancy. In addition, a similar work is required on the role and influence of the external environmental factors. Weeds are of most concern since possess different types of dormancy and challenges farmers as well as researchers under field conditions. Literature on the causal factors of dormancy are huge and varied including information on the seed coat and its structures, effects of temperature, light and phytochrome system, hormones, synthetic chemicals, enzymes, temperature, O_2, CO_2, seed internal structures, embryo and its surrounding structures, inhibitors, cell membranes, secondary metabolites and allelochemicals, stresses, agricultural practices and genetics, soil moisture and relative humidity, salinity, soil pH, and many other related factors. These factors and their interactions influence seed dormancy and germination. The interaction between environmental factors and seed factors that determines seed germination time, periodicity, sequences and percentages and the final density of the emerged weed population. However, while researchers all over the world are trying to solve and reveal the secrecy stands behind seed dormancy in order to find solutions for some problems that the farmers facing under field conditions at which seed dormancy of weeds is the main issue, dormancy from the weedness perspectives is a survival strategy that these unwanted plant species adapt themselves to survive and exist free from hazards and insure their generations and genetic lines. Therefore dormancy is a natural phenomenon created through genetics, environment, or their interactions, while research work carried out till now is just to understand this trial and to accommodate ourselves accordingly with. Therefore, seed dormancy is one of the most important adaptive mechanisms in plants, which protects seeds from precocious germination in the presence of the inappropriate conditions for growth continuation.

Author details

Jamal R. Qasem
Department of Plant Protection, Faculty of Agriculture, University of Jordan, Amman, Jordan

*Address all correspondence to: jrqasem@ju.edu.jo

References

[1] Islam AKMM, Yeasmin S, Qasem JRS, Juraimi AS, Anwar MP. Allelopathy of medicinal plants: Current status and future prospects in weed management. Agricultural Sciences. 2018;**9**:1569-1588. DOI: 10.4236/as.2018.912110

[2] Grundy AC. Predicting weed emergence: A review of approaches and future challenges. Weed Research. 2003;**43**:1-11

[3] Dyer WE. Exploiting weed seed dormancy and germination requirements through agronomic practices. Weed Science. 2017;**43**(3):498-503. DOI: 10.1017/S0043174500081534

[4] Schatral A, Osborne JM. Germination and dormancy states of seeds in the environmental weed *Rumex vesicarius* (Polygonaceae). I. Seed polymorphism and germination of extracted seeds. Australian Journal of Botany. 2006;**54**(8):773-782. DOI: 10.1071/BT04201

[5] White DE. Physiological adaptations in two ecotypes of canada thistle *Circium arvense* (L.) Scop [MSc thesis]. Davis, USA: University of California; 1979

[6] Altieri MA, Liebman MZ. Weed Management in Agroecosystems: Ecological Approaches. Boca Raton: CRC Press; 1988

[7] Batlla D, Benech-Arn RL. Predicting changes in dormancy level in weed seed soil banks: Implications for weed management. Crop Protection. 2007;**26**:189-197

[8] Vleeshouwers LM. Modelling weed emergence patterns [Ph.D. thesis]. Wageningen, The Netherlands: Agricultural University; 1997

[9] Benech-Arnold RL, Sanchez RA, Forcella F, Kruk BC, Ghersa CM. Environmental control of dormancy in weed seed banks in soil. Field Crop Research. 2000;**67**:105-122

[10] Mayer AM, Poljakoff-Mayber. The Germination of Seeds. 3rd ed. Oxford: Pergamon Press; 1982. 211 pp

[11] Chavarria PL. Seed dormancy characteristics in six weed species as affected by after-ripening temperatures and field conditions. Retrospective theses and dissertations. 8062 [Ph.D thesis]. Ames, Iowa: Iowa State University; 1986. Available from: https://lib.dr.iastate.edu/rtd/8062

[12] Nikolaeva MG. Factors controlling the seed dormancy pattern. In: Khan AA, editor. The Physiology and Biochemistry of Seed and Germination. Amsterdam, The Netherlands: Elsevier/North-Holland Biomedical Press; 1977. pp. 51-74

[13] Venier P, Funes G, Garcia C. Physical dormancy and histological featuresof seeds of five Acacia species (Fabaceae) from xerophytic forests in central Argentina. Flora. 2012;**207**:39-46

[14] Gama-Arachchige N, Baskin J, Geneve R, Baskin C. Identification and characterization of ten new water gaps in seeds and fruits with physical dormancy and classification of water-gap complexes. Annals of Botany. 2013;**112**:69-84

[15] Black M, Bewley J, Halmer P. The Encyclopaedia of Seeds: Science, Technology and Uses. London: CABI; 2006

[16] Sm'ykal P, Vernoud V, Blair MW, Soukup A, Thompson RD. The role of the testa during development and in establishment of dormancy of the legume seed. Frontiers in Plant Science. 2014;**5**:1-19

[17] Rodrigues-Junior A, Faria J, Vaz T, Nakamura A, Jose A. Physical

dormancy in *Senna multijuga* (Fabaceae: Caesalpinioideae) seeds: The role of seed structures in water uptake. Seed Science Research. 2014;**24**:147-157

[18] Cardina J, Sparrow DH. Temporal changes in velvetleaf (*Abutilon theophrasti*) seed dormancy. Weed Science. 2017;**45**(1):61-66. DOI: 10.1017/S0043174500092481

[19] Robles-Díaza E, Floresb J, Yáñez-Es, L. Paths of water entry and structures involved in the breaking of seed dormancy of Lupinus. Journal of Plant Physiology. 2016;**192**:75-80

[20] Brown NAC, Bridglall SS. Preliminary studies of seed dormancy in *Datura stramonium*. Souh Africa Journal of Botany. 1987;**53**(1):107-109

[21] Qasem JR. Aqueous extract effect of nettle-leaved goosefoot (*Chenopodium murale*) on wheat and barley. Research Journal of Aleppo University. 1990;**14**:37-53

[22] Malavert C, Batllaa D, Benech-Arnolda RL. Temperature-dependent regulation of induction into secondary dormancy of *Polygonum aviculare* L. seeds: A quantitative analysis. Ecological Modelling. 2017;**352**:128-138

[23] Karlsson LM, Milberg P. Seed dormancy pattern and germination preferences of the South African annual *Papaver aculeatum*. South African Journal of Botany. 2007;**73**:422-428

[24] Dahlquist RM, Prather TS, Stapleton JJ. Time and temperature requirements for weed seed thermal death. Weed Science. 2007;**55**:619-625

[25] Simpson GM. Seed Dormancy in Grasses. Cambridge, UK: Cambridge University Press; 1990

[26] Carrera E, Holman T, Medhurst A, Dietrich D, Footitt S, Theodoulou FL, et al. Seed after-ripening is a discrete developmental pathway associated with specific gene networks in Arabidopsis. Plant Journal. 2008;**53**:214-224

[27] Finch-Savage WE, Leubner-Metzger G. Seed dormancy and the control of germination. New Phytologist. 2006;**171**:501-523

[28] Koornneef M, Karssen CM. Seed dormancy and germination. In: Koornneef M, Karssen CM, editors. Arabidopsis. Cold Spring Harbor, NY: Cold Spring Harbor Laboratory Press; 1994. pp. 313-334

[29] Leopold AC, Glenister R, Cohn MA. Relationship between water content and afterripening in red rice. Physiologia Plantarum. 1988;**74**:659-662

[30] Kołodziejek J, Patykowski J. The effect of temperature, light and calcium carbonate on seed germination and radicle growth of the polycarpic perennial *Galium Cracoviense* (Rubiaceae), a narrow endemic species from southern Poland. Acta Biologica Cracoviensia Series Botanica. 2015;**57**(1):70-81. DOI: 10.1515/abcsb-2015-0006

[31] Karssen CM. Seasonal patterns of dormancy in weeds seeds. In: Khan AA, editor. The Physiology and Biochemistry of Seed Development, Dormancy and Gennination. Amsterdam, The Netherlands: Elsevier/North-Holland Biomedical Press; 1982. pp. 243-270

[32] Egley GH. Seed germination in soil: Dormancy cycles. In: Kigel J, Galili J, editors. Seed Development and Germination. New York: Marcel Dekker, Inc.; 1995. p. 853

[33] Hillhorst HWM. The regulation of secondary dormancy. The membrane hypothesis revised. Seed Science Research. 1998;**8**:77-90

[34] Taab A. Seed dormancy and germination in solanum nigrum

and *S. physalifolium* as influenced by temperature conditions [Doctoral thesis]. Uppsala: Swedish University of Agricultural Sciences; 2009. 48 pp

[35] Kołodziejek J, Patykowski J. Effect of environmental factors on germination and emergence of invasive *Rumex confertus* in Central Europe. The Scientific World Journal. 2015;**2015**:170176. DOI: 10.1155/2015/170176

[36] Arc E, Sechet J, Corbineau F, Rajjou L, Marion-Poll A. ABA cross talk with ethylene and nitric oxide in seed dormancy and germination. Frontiers in Plant Science, Plant Cell Biology. 2013;**4**:1-19. DOI: 10.3389/fpls.2013.00063

[37] Karssen CM. Two sites of hormonal action during germination of *Chenopodium album* seeds. Physiologia Plantarum. 1976;**36**:264-270

[38] Rezvani M, Zaefarian F, Amini V. Effects of chemical treatments and environmental factors on seed dormancy and germination of shepherd's purse (*Capsella bursa-pastoris* (L.) Medic.). Acta Botanica Brasilica. 2014;**28**(4):495-501. DOI: 10.1590/0102-33062014abb3337

[39] Rice EL. Allelopathy. 11th ed. London: Academic Press; 1984

[40] Bouteau HE-M, Bailly C. Oxidative signaling in seed germination and dormancy. Plant Signaling & Behavior. 2008;**3**(3):175-182. DOI: 10.4161/psb.3.3.5539

[41] Finch-Savage WE, Footitt S. Seed dormancy cycling and the regulation of dormancy mechanisms to time germination in variable field environments. Journal of Experimental Botany. 2017;**68**(4):843-856. DOI: 10.1093/jxb/erw477

[42] Koller D. Environmental control of seed germination. In: Kozlowski TT, editor. Seed Biology. Vol. II. New York: Academic Press; 1972. pp. 1-101

[43] Menegat A, Milberg P, Nilsson ATS, Andersson L, Vico G. Soil water potential and temperature sum during reproductive growth control seed dormancy in *Alopecurus myosuroides* Huds. Ecology and Evolution. 2018;**8**:7186-7194

[44] Menge EO, Bellairs SM, Lawes MJ. Seed-germination responses of *Calotropis procera* (Asclepiadaceae) to temperature and water stress in northern Australia. Australian Journal of Botany. 2016;**64**(5):441-450. DOI: 10.1071/BT16044

[45] Xiong RC, Ma Y, Wu HW, Jiang WL, Ma XY. Effects of environmental factors on seed germination and emergence of velvetleaf (*Abutilon theophrasti*). Planta Daninha. 2018;**36**. Epub Oct 29, 2018. DOI: 10.1590/s0100-83582018360100122

[46] Bochenek A, Golaszewski J, Piotrowicz-Cieslak AI, Gorecki RJ. The effects of temperature on the dormancy and germination of *Cirsium arvense* (L.) Scop. seeds. Acta Societatis Botanicorum Poloniae. 2009;**78**(2):105-114

[47] Fei H, Tsang E, Cutler AJ. Gene expression during seed maturation in Brassica napus in relation to the induction of secondary dormancy. Genomics. 2007;**89**:419-428

[48] Cici SZH. Maternal environment modulates dormancy and germination in *Vaccaria hispanica*. Agrotechnology. 2017;**6**:154. DOI: 10.4172/2168-9881.1000154

[49] Humphries T, Chauhan BS, Florentine SK. Environmental factors effecting the germination and seedling emergence of two populations of an aggressive agricultural weed; *Nassella trichotoma*. PLoS One. 2018;**13**(7):e0199491. DOI: 10.1371/journal.pone.0199491

[50] Orlovsky N, Japakova U, Zhang H, Volis H. Effect of salinity on seed germination, growth and ion content in dimorphic seeds of Salicornia europaea L. (Chenopodiaceae). Plant Diversity. 2016;**38**:183-189

[51] Soltani E, Baskin CC, Baskin JM, Soltani A, Galeshi S, Ghaderi-far F, et al. A quantitative analysis of seed dormancy and germination in the winter annual weed *Sinapis arvensis* (Brassicaceae). Botany. 2016;**94**(4):289-300. DOI: 10.1139/cjb-2015-0166

[52] Teasdale JR, Pillai P. Contribution of ammonium to stimulation of smooth pigweed (*Amaranthus hybridus* L.) germination by extracts of hairy vetch (*Vicia villosa* Roth) residue. Weed Biology and Management. 2005;**5**:19-25. DOI: 10.1111/j.1445-6664.2005.00155.x

[53] Brainard DC, Tommaso AD, Mohler CL. Intraspecific variation in germination response to ammonium nitrate of *Amaranthus powellii* originating from organic versus conventional vegetable farms. Weed Science. 2006;**54**:435-442. DOI: 10.1614/WS-05-162R1.1

[54] Qasem JR. Chemical control of *Prosopis farcta* (Banks and Sol.) Macbirde in the Jordan Valley. Crop Protection. 2007;**26**:572-575

[55] Vleeshouwers L, Bouwmeester H, Karssen C. Redefining seed dormancy: An attempt to integrate physiology and ecology. Journal of Ecology. 1995;**83**:1031-1037

[56] Hawkins KK, Allen PS, Meyer SE. Secondary dormancy induction and release in *Bromus tectorum* seeds: The role of temperature, water potential and hydrothermal time. Seed Science Research. 2017;**27**:1-14. DOI: 10.1017/S0960258516000258

[57] Putwain PD, Scott KR, Holliday RJ. The nature of resistance to traizine herbicides: Case histories of phenology and population studies. In: Le Baron HM, Gressel J, editors. Herbicide Resistance in Plants. New York: John Wiley & Sons; 1982. p. 99

[58] Harper JL. Population Biology of Plants. London: Academic Press; 1977

[59] Bewley JD, Black M. Physiology and Biochemistry of Seeds in Relation to Germination, Viability, Dormancy and Environmental Control. Vol. 2. New York: Springer-Verlag; 1982. 375 pp

[60] Abbaspour R, Cici SZH, Dogaheh MA, Abbaspour N. Effect of fertilizer application and irrigation on germination of Jimson weed (*Datura stramonium* L.) as a medicinal plant. International Journal of Medical Research & Health Sciences. 2016;**5**(S):377-380

[61] Boddy LG, Bradford KJ, Fischer AJ. Stratification requirements for seed dormancy alleviation in a wetland weed. PLoS One. 2013;**8**(9):e71457. DOI: 10.1371/journal.pone.0071457

[62] Scursoni JA, Benech-Arnold RL, Hirchoren H. Demography of wild oat in barley crops: Effect of crop, sowing rate and herbicide treatment. Agronomy Journal. 1999;**91**:478-485

[63] Ali HH, Tanveer A, Nadeem MA, Asghar HN. Methods to break seed dormancy of *Rhynchosia capitata*, a summer annual weed. Chilean Journal of Agricultural Research. 2011;**71**(3):483-487

[64] Chauhan BS, Johnson DE. Influence of environmental factors on seed germination and seedling emergence of Eclipta (Eclipta prostrate) in a tropical environment. Weed Science. 2008;**56**:383-388

[65] Tiryaki I, Topu M. A novel method to overcome coat-imposed seed

dormancy in *Lupinus albus* L. and *Trifolium pratense* L. Journal of Botany. 2014;**2014**, Article ID 647469, 6 pages. DOI: 10.1155/2014/647469

[66] Wu H, Zhang P, Chen G, Pan L, Li J, Dong L. Environmental factors on seed germination and seedling emergence of *Phleum paniculatum* Huds. Chilean Journal of Agricultural Research. 2018;**78**(3):370-377. DOI: 10.4067/S0718-58392018000300370

[67] Karlsson LM, Milberg P. Variation within species and inter-species comparison of seed dormancy and germination of four annual *Lamium* species. Flora. 2008;**203**:409-420

[68] Bouwmeester HJ, Karssen CM. Annual changes in dormancy and germination in seeds of *Sisymbrium officinale* (L.). Scop. New Phytologist. 1993;**124**:179-191

[69] Hoyle GL, Cordiner H, Good RB, Nicotra AB. Effects of reduced winter duration on seed dormancy and germination in six populations of the alpine herb *Aciphyllya glacialis* (Apiaceae). Conservation Physiology. 2014;**2**:1-11

[70] Rhie YH, Kim J, Lee SY, Kim KS. Non-deep simple morphophysiological dormancy in seeds of heavenly bamboo (*Nandina domestica* Thunb.). Scientia Horticulturae. 2016;**210**:180-187

[71] Skubacz A, Daszkowska-Golec A. Seed Dormancy: The Complex Process Regulated by Abscisic Acid, Gibberellins, and Other Phytohormones that Makes Seed Germination Work. IntechOpen; 2017. DOI: 10.5772/intechopen.68735. Available from: https://www.intechopen.com/books/phytohormones-signaling-mechanisms-and-crosstalk-in-plant-development-and-stress-responses/seed-dormancy-the-complex-process-regulated-by-abscisic-acid-gibberellins-and-other-phytohormones-th

[72] Riemens MM, Scheepens PC, van der Weide RY. Dormancy, Germination and Emergence of Weed Seeds, with Emphasis on the Influence of Light. Vol. 2004. Wageningen: Plant Research International B.V.; 2004. pp. 1-26

[73] Bouwmeester HJ. The Effect of Environmental Conditions on the Seasonal Dormancy Pattern and Germination of Weed Seeds. Wageningen, The Netherlands: Agricultural University in Wageningen, Department of Plant Physiology, Arboretumlaan 4, NL-6703 BD Wageningen, The Netherlands and Department of Vegetation Science, Plant Ecology and Weed Science; 1990

[74] Karimmojeni H, Taab A, Rashidi B, Bazrafshan AH. Dormancy breaking and seed germination of the annual weeds *Thlaspi arvense*, *Descurainia sophia* and *Malcolmia africana* (Brassicaceae). Journal of Plant Protection Research. 2014;**54**(2):179-187

[75] Brainard D, Mohler C, Schonbeck M. Manipulating weed seed banks to promote their decline. Extension; Organic Agriculture. 2013. Available at: https://articles.extension.org/pages/18528/manipulating-weed-seed-banks-to-promote-their-decline [Accessed: May 15, 2019]

[76] Sánchez RA, de Miguel L. Phytochrome promotion of mannan-degrading enzyme activities in the micropylar endosperm of *Datura ferox* seeds requires the presence of the embryo and gibberellins synthesis. Seed Science Research. 1997;7:27-33

[77] Mohler CL. Weed life history: Identifying vulnerabilities. In: Liebman M et al., editors. Ecological Management of Agricultural Weeds. New York: Cambridge University Press; 2001. pp. 40-98

[78] Martens K, Martens M. Weed control in row crops. In: Oral

Presentation, Pennsylvania Association for Sustainable Agriculture Conference; Feb. 7-9 2008; State College, PA. 2008

[79] Davis A, Liebman M. Cropping systems effects on giant foxtail (*Setaria faberi*) demography: 1. Green manure and tillage timing. Weed Science. 2003;**51**:919-929. DOI: 10.1614/P2002-133A

[80] Gallagher RS, Granger KL, Snyder Am, Pittmann D, Fuerst EP. Implications of environmental stress during seed development on reproductive and seed bank persistence traits in wild oat (*Avena fatua* L.). Agronomy. 2013;**3**:537-549. DOI: 10.3390/agronomy3030537

[81] Dixon K. Releasing agricultural weed seed dormancy through application of a novel smoke derived chemical. In: Final Report. Botanic Garden and Parks Authority, West: Grains Research & Development Corporation, Australian Government; 2005

[82] Gorecki MJ, Long RL, Flematti GR, Stevens JC. Parental environment changes the dormancy state and karrikinolide response of Brassica tournefortii seeds. Annals of Botany. 2012;**109**:1369-1378. DOI: 10.1093/aob/mcs067. Available online at: www.aob.oxfordjournals.org

[83] Qasem JR. Parasitic weeds and allelopathy, from the hypothesis to the proof. In: Reigosa MJ, Pedrol N, Gonzalez L, editors. Allelopathy, A Physiological Process with Ecological Implications. The Netherlands: Springer; 2006. pp. 565-637

[84] Zwanenburg B, Reizelman A. En route to the isolation and characterization of the strigolactone receptor using biotin labelled strigolactone analogues. In: Fer A, Thalouam P, Joel DM, Musselman LI, ParkerC, VerkleijJAC, editors. Proceedings of the 7th International Parasitic Weed Symposium. Nantes, France: Faculté des Sciences, Université de Nantes; 2001. pp. 102-105

[85] Butler LG. Chemical communication between the parasitic weed Striga and its crop host. A new dimension in allelochemistry. In: Inderjit, Dakshini KMM, Einhellig FA, editors. Allelopathy: Organisms. Processes and Applications. Washington, DC: ACS Symposium Series 582. American Chemical Society; 1995. pp. 389. pp. 158-168

[86] Denev I, Pereira A, Verstappen F and Bouwmeester H. Biosynthesis of *Orobanche* germination stimulants. In Fer A, Thalouarn P, Joel DM, Musselman LJ, Parker C, Verkleij JAC (eds) Proceedings of the 7th International Parasitic Weed Symposium. 2001. pp. 110-113. Nantes, France: Faculté des Sciences, Université de Nantes; DOI: 10.2478/jppr-2014-0028

[87] Sugimoto Y. Germination stimulants for the seeds of root parasitic weeds. Journal of Pesticide Science. 2000;**25**:438-440

[88] Yokota T, Sakai H, Okuno K, Yoneyama K, Takeuchi Y. Alectrol and orobanchol, gennination stimulants for *Orobanche minor*, from its host red clover. Phytochemistry. 1998;**49**:1967-1973

[89] Yoneyama K, Takeuchi Y, Yokota T. Natural gennination stimulants for Orobanche minor. In: Fer A, Thalouam P, Joel DM, Musselman LI, ParkerC, VerkleijJAC, editors. Proceedings of the 7th International Parasitic Weed Symposium. Nantes, France; 2001. p. 123

[90] Wegmann K. The *Orobanche* problem in tobacco. In: Wegmann K, Musselman LJ, Joel DM, editors. Current Problems of Orobanche Research. Proceedings of the Fourth International

Workshop on Orobanche. Albena, Bulgaria; 1998. pp. 21-24

[91] Perez-de-Luque A, Rubiales D, Galindo GC, Macias FA, Jorrin J. Allelopathy and allelochemicals within the plant-parasitic weed interaction. Studies with the *sunflower-Orobanche cumana* system. In: Fer A, Thalouam P, Joel DM, Musselman LI, ParkerC, VerkleijJAC, editors. Proceedings of the 7'hInternational Parasitic Weed Symposium. France: Nantes; 2001. pp. 196-199

[92] Zehar N, Fer A. Gennination of Orobanche ramosa is controlled by gibberellins and ethylene. In: Fer A, Thalouam P, Joel DM, Musselman L, ParkerC, VerkleijJAC, editors. Proceedings of the 7th International Parasitic Weed Symposium. Nantes, France: Faculté des Sciences, Université de Nantes; 2001. p. 122

[93] Worsham AD, Egley GH. Physiology of witchweed seed donnancy and gennination. In: Sand PF, Eplee RE, Westbrooks RG, editors. Witchweed Research and Control in the United States. Champaign, USA: Weed Science Society of America; 1990. pp. 11-26

[94] Babiker AGT, Parker C, Suttle JC. Induction of *Striga* seed germination by thidiazuron. Weed Research. 1992;**32**:243-248

[95] Egley GH. Ethylene stimulation of weed seed germination. Agriculture and Forestry Bulletin, University of Alberta. 1982;**5**:13-18

[96] Thuring JWFW, Heinsman NWJT, Jacobs RWAWM, Nefkens GHL, Zwanenburg B. Asymmetric synthesis of all stereoisomers of demethylsorgolactone. Dependence of the stimulatory activity of *Striga hermonthica* and *Orobanche crenata* seed gennination on the absolute configuration. Journal of Agriculture and Food Chemistry. 1997;**45**:507-513

[97] Wigchert SCM, Kuiper E, Boelhouwer G, Nefkense GHL, Verkleij JA, Zwanenburg B. Dose-response of seeds of the parasitic weeds *Striga* and *Orobanche* toward the synthetic germination stimulants GR24 and Nijmegen I. Journal of Agriculture and Food Chemistry. 1999;**47**:1705-1710

[98] Yoneyama K, Ogasawara M, Takeuchi Y, Konnal M, Sugimoto Y, Sassa T, et al. Effect of jasmonates and related compounds on seed germination of *Orobanche minor* Smith and *Striga hermonthica* (Del.) Benth. Bioscience, Biotechnology and Biochemistry. 1998;**62**:1448-1450

[99] Vail SL, Dailey OO Jr, Connik W, Peppennan AB. Strigol synthesis and related structurebioactivity studies. In: Thompson A, editor. The Chemistry of Allelopathy. Biochemical Interactions among Plants, American Chemical Society Symposium Series 268. Washington, DC, USA: American Chemical Society; 1985. pp. 445-456

[100] deLuqueAP, GalindoJCG, MaciasFA, Jorrin J. Sunflower sesquiterpene lactone models induce *Orobanche cumana* seed germination. Phytochemistry. 2000;**53**:45-50

[101] Takeuchi Y, Omigawa Y, Ogasawara M, Yoneyama K, Konnai M, Wosham AD. Effects of brassinosteroids on conditioning and germination of clover broomrape (*Orobanche minor*) seeds. Plant Growth Regulator. 1995;**16**:153-160

[102] Christeva T, Naumova S. Stimulation of broom rape seed germination by soil microorganisms. In: Wegmann K, Musselman LJ, Joel DM, editors. Current Problems of Orobanche Researches. Proceedings of the Fourth International Workshop on Orobanche. Albena, Bulgaria: Institute for wheat and sunflower "Dobroudja"; 1998. pp. 73-77

[103] Yoneyama K, Takeuchi Y, Ogasawara M, Konnal M, Sugimoto Y, Sassa T. Cotylenins and fusicoccins stimulate seed germination of *Striga hermonthica* (Del.) Benth and *Orobanche minor* Smith. Journal of Agriculture and Food Chemistry. 1998;**46**:1583-1586

[104] Babaei S, Alizadeh H, Jahansouz MR, Mashhadi HR, Moeini MM. Management of *Phelipanche aegyptiaca* Pomel. using trap crops in rotation with tomato (*Solanum lycopersicon* L.). Australian Journal of Crop Science (AJCS). 2010;**4**(6):437-442

[105] Ma YQ, Zhang W, Dong SQ, Ren XX, An Y, Lang M. Induction of seed germination in *Orobanche* spp. by extracts of traditional Chinese medicinal herbs. Science China Life Sciences. 2012;**55**:250-260. DOI: 10.1007/s11427-012-4302-2

[106] Qasem JR. Branched broomrape (*Orobanche ramosa* L.) control in tomato (*Lycopersicon esculentum* Mill.) by trap crops and other plant species in rotation. Crop Protection. 2019;**120**:75-83

[107] Serghini K, Perez de Luque A, Castejon-Munoz M, Garcia-Torres L, Jorrin JV. Sunflower (*Helianthus annuus* L.) response to broomrape (*Orobanche cernua* Loefl.) parasitism: Induced synthesis and excretion of 7-hydroxylated simple coumarins. Journal of Experimental Botany. 2001;**52**:2227-2234

[108] Guitierrez-Mellado MC, Edwards R, Tena M, Cabello F, Serghini K, Jorrin J. The production of coumarin phytoalexins in different plant organs of sunflower (*Helianthus annuus* L.). Journal of Plant Physiology. 1995;**149**:261-266

[109] Bar-Nun N, Mayer AM. Composition of and changes in storage compounds in *Orobanche. aegyptiaca* seeds during preconditioning. Journal of Plant Sciences. 2002;**50**:277-279

[110] Sauerborn J, Buschmann H, Ghiasi KG, Kogel KH. Benzothiadiazole activates resistance in sunflower (*Helianthus annuus*) to the root-parasitic weed *Orobanche cumana*. Phytopathology. 2002;**92**:59-64

[111] Sauerborn J. Angiospennous root parasites and fungi as natural antagonists. In: Fer A, Thalouam P, Joel DM, Musselman LI, Parker C, Verkleij JAC, editors. Proceedings of the 7th International Parasitic Weed Symposium. Nantes, France: Faculté des Sciences, Université de Nantes; 2001. pp. 248-251

[112] Strobel GA, Kenfield D, Bunkers Q, Sugawara F, Clardy J. Phytotoxins as potential herbicides. Experientia. 1991;**47**:819-826

[113] Zonno MC, Vurro M. Effect of fungal toxins on germination of *Striga hermonthica* seeds. Weed Research. 1999;**39**:15-20

[114] Gabbar A, Babiker T, Buttler LG, Ejeta G, Woodson WR. Enhancement of ethylene biosynthesis and germination by cytokinins and 1-aminocyclopropane-I-carboxylic acid in *Striga asiatica* seeds. Physiologia Plantarum. 1993;**89**:21-26

[115] Anderson JA, Sorrells ME, Tanksley SD. RFLP analysis of genomic regions associated to pre-harvest sprouting in wheat. Crop Science. 1993;**33**:453-459

[116] Paterson AH, Sorrells ME. Inheritance of grain Upadhyay, M. P., and G. M. Paulsen, 1988. Heritabilities and gedormancy in white-kernelled wheat. Crop Science. 1990;**30**:25-30

[117] Upadhyay MP, Paulsen GM. Heritabilities and dormancy in white-kernelled wheat. Crop Science. 1988;**30**:25-30

[118] Groos C, Gay G, Perretant MR, Gervais L, Bernard M, et al. Study

of the relationship between pre-harvest sprouting and grain color by quantitative trait loci analysis in a white x red grain bread-wheat cross. Theoretical and Applied Genetics. 2002;**104**:39-47

[119] Gu X-YG, Kianian SF, Foley ME. Multiple loci and epistases control genetic variation for seed dormancy in weedy rice (*Oryza sativa*). Genetics. 2004;**166**:1503-1516

[120] Fenner M. Seed Ecology. New York: Chapman and Hall; 1985. 151 pp

[121] Tran VN, Cavanagh AK. Structural aspects of dormancy. In: Murray DR, editor. Seed Physiology, Germination and Reserve Mobilization. Vol. II. Sydney, Australia: Academic Press; 1984. pp. 1-44

[122] Roberts EH. The interaction of environmental factors controlling loss of dormancy in seeds. Annals of Applied Biology. 1981;**98**:552-555

Chapter 3

Artificial Intelligence Tools to Better Understand Seed Dormancy and Germination

Manuel Ayuso, Mariana Landín, Pedro Pablo Gallego and Mª Esther Barreal

Abstract

Despite a large number of publications available, the control mechanisms of seed dormancy and germination are far to be fully understood. Seed dormancy and germination are very complex biological processes and because they involve multiple factors (physiological, mechanical, and environmental) and their nonlinear interactions. This explains why extremely little variations on some of those factors and in the way they interact caused enormous variation in the obtained results. Multifactorial process like these can be modeled using computer-based tools to predict better results. In this chapter, some basic concepts relative to seed dormancy and germination and the main factors (physiological, involved in seed dormancy, particularly dormancy-inducers and dormancy-breakers, and seed germination) are reviewed. In the next two, we describe the use of artificial intelligence computer-based models to better understand the physiological mechanisms of seed dormancy (how dormancy is controlled and how can be released) and seed germination. Finally, some applications of artificial neural networks, fuzzy logic, and genetic algorithms to elucidate critical factors and predict optimal condition for seed dormancy-breaking and germination are given as examples of the utility of this powerful computer-based tools.

Keywords: primary dormancy, secondary dormancy, dormancy maintenance, dormancy release, germination factors, temperature fluctuations, light exposure, moisture, day length, after-ripening, stratification

1. Introduction

The importance of the seeds began with the dawn of agriculture, around 12,000 years ago, although seeds have been collected and eaten for many thousands of years before crop domestication (20,000–100,000 years). This domestication involved the selection of the desirable traits, as a high yield, appropriated seed size and good resistance/tolerance to biotic and abiotic stress, avoiding undesirable ones (mechanism of dispersion and seed latency).

The knowledge about the storage, distribution, germination, sowing, and harvest of seeds improved for the following centuries. The first written references on the germination of seeds can be found in religious texts or in the "naturalist-texts of Greeks and Romans." In those documents, Theophrastus and Pliny, the elder explain various germination concerns, as the need of drying the seeds for storage or soaking them in water or in milk to stimulate their germination [1].

Germination is a complex physiologic process, beginning with water imbibition by the seeds and ending with the emergence of one part of the embryonic organ, the radicle. Harvested mature seeds are usually quiescent, meaning that they may survive many years with a standstill metabolism and low water content (<15%). Quiescent seeds must be imbibed for being able to activate its metabolism and germinate under suitable environmental conditions [2]. During seed imbibition, water uptake triggers the resumption of seed normal metabolic levels, and promote the damage repair occurred during drying. Once the seeds return to their normal metabolic state, an expansion of embryonic cells causes the embryo emergence and marks the end of germination. However, some imbibed and metabolically active seeds cannot germinate under a wide range of normal environmental factors and hence, they are considered dormant [3].

Seed dormancy plays a key role in the regulation of germination [4, 5]. Dormancy induction, maintenance, and release are determined by physiological and morphological seed characteristics and their control are governed by many genetic and environmental factors. Dormancy is then, a very complex biological process that involves multiple interactive factors (physiological, mechanical, and environmental), making it difficult to fully understand its performance despite the large number of publications available. Therefore, understanding how dormancy can be controlled and/or released should ensure the success of germination in desirable species with very interesting consequences in the socio-economic and research fields [4].

The factors that affect dormancy release and germination are generally studied independently, although they are obviously interconnected: a) no germination is being possible without dormancy-breaking, b) it is almost impossible to define the beginning and the end of each process and, c) many factors may interact or counteract in both processes.

Traditionally, data from dormancy or germination studies were analyzed using traditional statistical methods, nevertheless, complex biological process such as germination and seed dormancy, cannot be fully understand by simple comparison of means among treatments, analysis of variance, regression models or simple algorithms, with those approaches being necessary to integrate multidimensional data to describe complex biological interactions [6, 7].

Artificial intelligence (AI) tools have been shown as useful techniques for establishing relationships between multiple variables (factors and parameters) [8–10]. In addition, several studies have shown the effectiveness of those AI tools, such as artificial neural networks (ANNs) combined with fuzzy logic or genetic algorithms for modeling and optimizing complex biological processes [11, 12].

In this chapter, we describe how AI models can be used for a better understanding and selection of the critical factors that stimulate the physiological mechanism of dormancy-breaking and germination in seeds.

2. Seed dormancy and germination: some basic concepts

Dormancy is an evolutionary characteristic that has increased the survival of plant species, through the inhibition of seed germination in adverse conditions [13]. Seed dormancy could be considered a germination absence under suitable environmental conditions, in an intact viable seed. This germination lack has evolved differently across the species and hence, several dormancy mechanisms have been developed according to the diversity of climates and habitats [5].

Nikolaeva developed the first dormancy classification scheme including just two kinds of dormancy: endogenous and exogenous [14]. In the first one, embryo prevents germination, while in the second one, some seed structures or chemicals are responsible for germination inhibition [14]. Later, Bewley and collaborators described the mechanism involved in these two dormancies [2, 15]. According to these authors, the endogenous dormancy, re-named as embryo dormancy, can be induced by undifferentiated embryos, immature embryos, chemical inhibitors (present in seeds) and physiological constraints. On the other side, exogenous dormancy, re-named by these authors as coat-imposed dormancy, is caused by several covering tissues that interfered with or suppress seed germination. These tissues inhibit water uptake, gas exchange, chemical inhibitors release or cause mechanical restraint. Plants can exhibit one or both types of dormancy, acting simultaneously or successively.

These different terms and definitions of seed dormancy have caused confusion within the scientific community because they include both morphological and physiological properties of the seeds. In the earliest years of this century, Baskin and Baskin [3] proposed a comprehensive classification, which accurately reflects all the points of view mentioned above, which comprise five main classes of seed dormancy: physiological (PD), morphological (MD), morphophysiological (MPD), physical (PY) and combinational (PY + PD).

Briefly, in PD class, seeds are water-permeable but present a physiological mechanism in the embryo that inhibits seed germination. Seeds with PD are affected by the phytohormone abscisic acid (ABA) and their physiological-inhibition grade (deep, intermediate and nondeep) varies according to their response to other phytohormone gibberellic acid (GA) and the breaking-dormancy requirements. Seeds showing MD dormancy have a small immature (underdeveloped, but differentiated) embryo and therefore, just needing an extra incubation time and suitable conditions for normal embryo development and following germination. However, in the case of seeds belonging to MPD dormancy, in addition to presenting an underdeveloped embryo, they also show PD that should be first broken to allow full embryo development, through warm/cold stratification and /or GA treatments. Concerning PY, it is caused by one or several cell layers that avoid the entry of water in the seed and can be broken under natural (high and/or fluctuating temperatures, fire, drying, digestive animal tract transit, and so on) or artificial (chemical or mechanical scarification or abrasion) conditions. By the formation of a gap between these waterproof coats, letting the water available for the embryo, germination can be restored. Finally, the combination of PY and PD, make the breaking the waterproof layers and the embryo physiological dormancy, necessary for achieving germination.

Despite all the above, it is important to emphasize that PD is the widest widespread, prevalent and abundant dormancy class for seeds from gymnosperms and angiosperms [3–5]. The wide distribution of plants with PD seeds have triggered in the appearance of different physiological mechanisms of induction or maintenance of dormancy. Moreover, these mechanisms are related to the environmental characteristics where the mother plant has grown [4, 5]. Many species with PD seeds show a cyclic change in dormancy states (primary and secondary), governed by several factors (**Figure 1**).

Induction of primary dormancy may impose during seed development by endogenous factors and its function is to prevent precocious germination, while seeds are being developed in the mother plant or immediately after their dispersal [2]. The plant hormone ABA is the main endogenous factor involved in the primary dormancy induction; however, exogenous factors such as environmental factors have also a high influence on this induction. In this sense, under optimal or at least

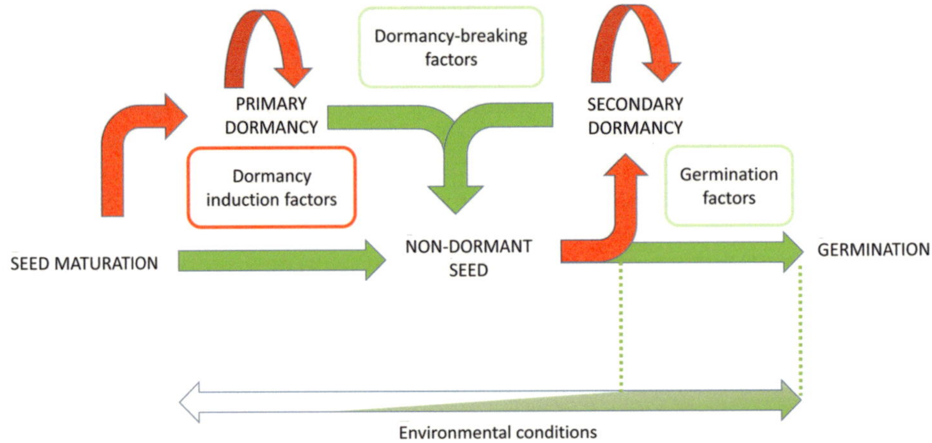

Figure 1.
Scheme of seed dormancy and germination control and regulation in response to environmental conditions. Several factors are involved on dormancy induction (red) and release (green arrows). Germination process can only be fully achieved if seed germination requirements (thresholds or sensitivity) overlap with adequate environmental conditions (green dotted area).

favorable conditions, the seeds do not suffer any germination block, moving to a nondormant state (**Figure 1**). Under favorable environmental conditions, germination starts with water uptake, followed by embryo expansion (embryo leave the dormant state, mobilizes stored nutrients, full elongate) and finish when breaks the covering coats and radicle protrusion occurs [4, 5, 16]. In any other circumstance, a blockage will happen at any time and the seeds will return to the dormant state.

Dormancy-breaking factors promote changes in dormant seeds (they cannot germinate in any condition) increasing their sensibility and allowing their germination under adequate environmental conditions [4, 5]. In addition, between dormancy and nondormant status, seeds are in a transitional state called conditional dormancy (CD), in which seeds are able to germinate but only in a small narrow of environmental conditions (**Figure 1**). In some species, under unfavorable conditions for germination, nondormant seeds may enter in a secondary dormancy status before germination [4, 17]. These seeds may continue in the transition between nondormant and dormant stage, which drives to a seasonal dormancy cycling (**Figure 1**). The cycle may continue for several years and is related to the maintenance of the soil seed bank, essential for the survival of plants communities.

The presence of nondormant and dormant seeds in a population depends on the effect of several induction factors during their development. These factors represent the main checkpoints in the control of germination and are summarized below.

2.1 Dormancy induction and maintenance factors

Seeds dormancy induction is highly correlated to the ABA content. This induction begins during seed development in the mother plant [18, 19]. In the initial stage, mother plant supplies ABA to seeds to prevent the premature germination [20–22]. In fact, Arabidopsis and maize mutants with reduced input of maternal ABA conduce to seeds germination in the mother plant [23, 24]. During the development, seeds begin to produce ABA by themselves and several scientific evidences demonstrated that this is the main factor required for inducing primary dormancy [5, 25, 26]. In fact, enhanced dormancy is evident in *Arabidopsis* mutants with overexpression of ABA biosynthesis genes, while ABA deficiency during seed development fails to induce primary dormancy. Therefore, it is widely accepted that the

ABA synthesized in the embryo and endosperm is the most critical factor inducing seed dormancy [16, 26].

ABA has also been proposed as the main factor involved in the dormancy maintenance of seeds, which still are dormant after their dispersion. In fact, *de novo* ABA biosynthesis has been associated to the maintenance of dormant state in several species (*Hordeum vulgare, Helianthus annuus*, and *Nicotiana plumbaginifolia*) [27–29]. In addition, *Arabidopsis thaliana* ecotype Cape Verde Island, imbibed seeds presented high ABA content and strong dormancy [17].

Recently, another phytohormone, the auxin indolacetic acid (IAA), has been found to regulate seed dormancy. Auxins have a similar effect than ABA, and transgenic seeds that overproduce auxin show a strong seed dormancy compared with wild-type seeds [30]. However, while the dormancy control of ABA by environmental signals is well-studied, the control of auxins in dormancy is not well-known yet [31].

The environment has a strong influence on the induction of dormancy during seed development (**Figure 1**). The main environmental factors affecting ABA content are temperature and light. They promote genetic expression changes in the mother plant during the seed development, which modify the ABA concentration and seed sensitivity to this phytohormone [2, 19]. Generally, low temperatures increase the dormancy induction during seed development in the mother plant, by increasing the expression of genes related to ABA biosynthesis, while high temperatures rise expression of ABA catabolism expression [2].

Dormancy induction is also influenced by natural light quantity (daily distribution), and quality (spectrum). Light daily distribution is controlled by the photoperiodic cycle: long days promote dormant seeds, while short days the opposite; the light quality is regulated via plant phytochrome (Prf): red wavelengths light (as white fluorescent) promote dormancy release, whereas far-red wavelengths (incandescent lights) promote dormancy induction and maintenance [32].

There are other factors, such as soil characteristics, that may also affect dormancy induction. Mineral nutrients such as nitrates, phosphate, sodium, potassium, zinc, iron, cooper taken up by the mother plant and translocated to the seeds has been included as dormancy inductors [2].

Finally, other mother plant physiological characteristics such as age, seed maturation timing, and seed position, which also can alter the dormancy induction [18].

2.2 Dormancy-breaking factors

The effect of ABA on dormancy induction and maintenance is counteracted by gibberellic acid (GA) since dormancy release depends mostly on ABA:GA balance (biosynthesis and catabolism) in seeds [33, 34]. This effect was demonstrated using plant deficient mutants. As example, Arabidopsis GA-deficient mutants present a strong seed dormancy and need exogenous application of GA for dormancy breaking [35], whereas mutants for genes involved in the negative regulators of GA biosynthesis pathway decreased the seed dormancy [36, 37]. Therefore, the release of latency depends on the concentration and sensitivity of the seeds to both phytohormones. Once seeds are imbibed, an increase in sensitivity and concentration of GA is necessary for dormancy release and some signals should trigger it. Nitric oxide (NO) has been proposed as a release dormancy signal since is related to the decrease of ABA sensibility and the increase of GA biosynthesis pathway in seeds of many species [38, 39]. Therefore, the dormancy release is established by the concentration and sensibility of both phytohormones in seeds.

Once seeds are released from the mother plant to the soil seed bank, they begin to behave as *sensors* which may detect environmental factors (signals) and change their dormancy status, affecting the expression of genes related to phytohormone metabolism [40]. The combination of environmental signals and phytohormonal metabolism provides a complex network that allows controlling the germination according to ecological opportunities [5, 26].

The dormancy-breaking mechanisms have been modified by evolution processes, according to the environmental signals in which the species lives. This facilitated that plant species with PD were dispersed and adapted to different habitats. Soil temperature and moisture are the major factors that indicate the seasonal changes. Both factors trigger the main modifications in the depth of dormancy, by changing the seeds sensibility to other environmental factors such as, light or nitrate among others [41].

Concerning those main factors, two main dormancy patterns have been found in the field. For species autumn-germinating, the dormancy is release during summer under warm and dry conditions, while for spring-germinating species the pattern shows a release of dormancy during winter under cold and wet conditions [42]. The evolutionary usefulness of dormancy is that the seeds need going through adverse germination conditions as a requirement to be able to break it and germinate conditions become appropriate. As example of this, has been described for seeds autumn-germinating that extend periods of warm temperature and dryness allowed release primary dormancy. This dormancy-breaking process is termed as after-ripening, promoting a decrease in ABA concentration in seeds, an increase GA sensitivity and widespread the range of other environmental requirements for germination. Ecologically, this requirement prevents germination during the hottest period of summer, being necessary for breaking dormancy and allowing the following germination in autumn.

The time required of after-ripening for release dormancy is highly genotype-dependent. Moreover, using different temperatures and moisture content the after-ripening may be accelerated [43]. In fact, the after-ripening improved the germination of three autumn-germinating species (*Anthocercis littorea*, *Dioscorea hastifolia*, and *Z. fruticulosum*) when the temperature and moist conditions were modified from their normal summer conditions [44].

As described above, seeds spring-germinating need periods of cold temperatures under moist conditions. This process is usually known as cold stratification or chilling. Seed stratification, promote the expression of genes related to GA biosynthesis and also decline the activity of some GA catabolic genes [45]. The stratification is required for majority nontropical species, which are spring germinating. Ecologically, this requirement prevents germination during their unfavorable season (winter) and allows their germination during spring, where suitable environmental conditions for seedling growth are settled (**Figure 1**). However, some species have a long period of cold stratification as requirement for break dormancy. In this case, a combination of after-ripening and cold stratification can be required in order to release dormancy [3].

Once the temperature and soil moisture have modified seed dormancy depth, seeds increase their sensibility to other breaking dormancy and germination factors. For example, light is considered an important environmental factor for releasing dormancy. Many species only break their dormancy by exposure to white light (i.e., sunlight), while other seeds only release dormancy by a change in photoperiod (i.e., length of the day).

Oxygen or carbon dioxide (soil gases) incorporated into the pores soil or dissolved in soil solution may affect the dormancy of the seeds. The seed responses to soil gases are highly variable and are dependent of the other environmental factors [2].

2.3 Germination factors

After dormancy release, nondormant seeds increase their sensitivity to the environmental factors favoring germination. To that end, GA stimulates germination in nondormant seeds by induction of hydrolytic enzymes, which stimulate the embryo growth, mobilization of endosperm storage reserves and weakening of tissues that are recovering the embryo [13]. Other phytohormones, such as ethylene and brassinosteroids, seems to be involved in some extension and limited impact on dormancy and germination, by reducing the influence of ABA effects in seeds [26, 46]. Indeed, exogenous applications of phytohormones as GA, cytokinins, or ethylene promoted germination in some species [47, 48].

Once phytohormones have induced seed sensibility, several environmental factors are involved on seed germination.

Temperature is a good seasonal indicator for seeds germination capacity and rate, although may induce secondary dormancy too. Usually, the temperature ranges for germinate are opposite to the ranges for release dormancy, as we described above. The range of temperature, in which seeds are able to germinate, falls into the next three categories: minimum, optimum and maximum. These ranges are related with the adaption of each species to their habitat and the favorable conditions for later seedling growth. As example, *Carex* sp. evidenced different temperature requirements for seed germination since they need a cold stratification for break dormancy but they are not able to germinate at low temperature. In fact, the best germination temperature was determined around 25°C [49]. Contrary, a study with 50 autumn-germinating species with after-ripening requirements, dormant or conditional dormant, demonstrated that they germinated only at low temperatures [50].

Water, particularly soil moisture, is an essential factor for seed germination. Water availability affects to the rate and speed of germination. The imbibition process, explained previously, allows the normal metabolic process resumption in nondormant seeds. In addition, it allows the radicle growth and elongation for break the seed coat [18].

Light is well known for stimulating germination in several species, since some nondormant hydrated seeds acquire high sensitivity to this factor after releasing their dormancy by after-ripening or chilling. The light requirement for germination prevents this process in unfavorable time or places for the seedling growth. Natural (fire) or cultivation (agricultural management) events caused soil disturbances, letting soil seed bank to be exposed to sunlight and favoring their germination [2].

Nitrate, nitric oxide and nitrites may stimulate the germination of many species. The ecological significance is that the seedling requires large amounts of nitrogen for optimal development. However, other germination factor/s may change the seed responsiveness to nitrate and their interaction regulates the germination response [51].

In conclusion, dormancy release and germination are sequential processes and it is too difficult to distinguish the end and the beginning of each one. It seems that seeds need opposite environmental conditions to germinate that those for release dormancy (i.e., temperature as explained above). In addition, the environmental requirements for germination are species-dependent. Therefore, the combinations of multiple factors (endogenous and environmental conditions) regulating seed dormancy and germination, makes so difficult to understand and predict the best germination conditions [2, 18, 37, 41].

3. Integrating information to understand dormancy and germination

Seed dormancy and germination are complex biological processes, as described previously. Understanding these processes and, subsequently predict and optimize them is a quite difficult task due to the high number of variables (factors) involved in dormancy and germination [7, 11]. In fact, using traditional methods, only a few number of factors can be studied simultaneously, so the fully understanding of dormancy and germination still remain a challenge.

Other limitation derived on the different kind of data generated by biological processes. Then the importance of each factor needs to be determined by diverse statistical analyses depending on data nature (binary, discrete, continuous, and so on). In addition, the traditional statistical analysis is quite limited to the possible interaction among factors and does not allow predicting their best factor conditions to optimize both processes. Therefore, for these kinds of processes, simple stepwise algorithms are useless and therefore, more complex analytical tools are required; such as multivariable approaches (networks) using computational models [52].

To develop a model some crucial steps are needed to be followed [7]:

1. Identify clearly the whole procedure (all steps) before built model.

2. Define and select the variables (factors and parameters) to study in the model.

3. Create a database with accurate data and select the kind of model to be built.

4. Validate the model to assure not significant differences among observed (experimental) and predicted (by the model) data.

Ishikawa diagram (**Figure 2**) is a useful chart to identify the causes of a specific event in order to fix some factors as controlled variables and select some as independent variables to be test experimentally [52]. This diagram shows the high

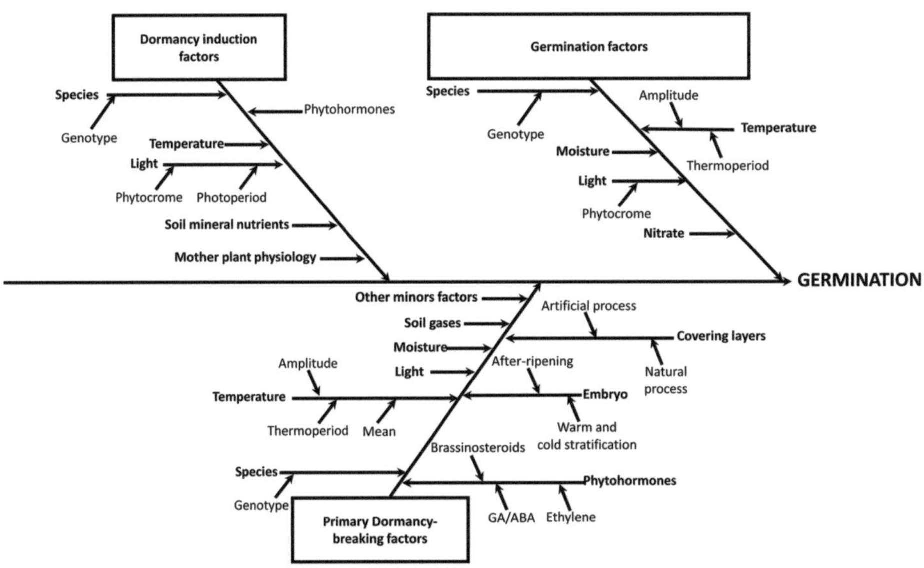

Figure 2.
Ishikawa diagram showing the large amount of factors (endogenous and exogenous) involved in seed dormancy (induction and breaking) and germination.

amount of factors and variables involved in the germination process and helps to identify some relationships among these factors overall process. After detecting the key variables (inputs) for germination, it is necessary to define the parameters for weight their effects (outputs) and an appropriate model. Although several models have been used to integrate data from complex biological processes, recent studies have shown the effectiveness of artificial intelligence tools as artificial neural networks (ANNs), genetic algorithms and neurofuzzy logic for modeling and optimize them [7, 12, 53, 54].

4. Artificial intelligence tools: from data to knowledge

Artificial Intelligence (AI) tools are problem-solving algorithms used when the number of solutions of one or several problems is huge, since they are capable of dealing with complex data in a versatile and powerful way [10]. This technology has been applied successfully to biomedical, pharmaceutical, and chemical applications for commercial and industrial purposes [8, 10, 55]. In addition, AI tools were used in basic and applied science research including ecology, environmental sciences, agriculture, food science, plant biology and biotechnology [9, 56, 57]. In this section, we summarized the most basic and important characteristics of the AI tools employed in plant biology [7].

4.1 Artificial neural networks (ANNs)

ANNs are AI tools that allow discovering nonlinear relationships among factors (inputs) and parameters (outputs) data. ANNs are one of the most effective method for revealing links among variables particularly if database are large and it is difficult to find direct relationships. In the case of germination (**Figure 2**), links between all factors and their effects on dormancy induction, dormancy-breaking and germination seems to be very complex, and difficult to be found using traditional statistical methodologies, but appropriate for ANNs. However, this technology has some limitations, related with the difficulties for results interpretation. In order to avoid this problem, ANNs are usually combined with other AI tools such as genetic algorithms or fuzzy logic (described below), which make easier the result interpretation [55, 58].

4.2 Genetic algorithms

Genetic algorithms (GA) are a heuristic algorithms based on genetic and natural selection. They are considered heuristic because generate useful solutions against problems. Concerning to plant biology, these algorithms have been used for optimization bioprocess [59]. Hybrids models between ANNs and GA have been performed allowing developing more accurate models for predict, optimize and control some biological process. Models achieved with this AI tools allowed determining the combinations of variables (factors or inputs) that provide the best results [7].

4.3 Fuzzy logic

Fuzzy logic tool is a computational tool, which allows analyzing and making deductions from uncertain or fuzzy data. Fuzzy logic assigns qualitative values using linguistic terms and degrees of membership called membership functions [7]. Interaction between linguistic terms (as low, medium or high) and membership functions allow the computer making meaning values to study [59]. Moreover, this tool explains the behavior of a complex system using an easy language and improves description about any complex task or process. Knowledge acquired after modeling with a fuzzy logic-based system, is expressed by IF-THEN rules. These rules explain antecedent conditions of inputs variables (factors studied) and their consequent effect on the output variables (parameters measured) [7]. In conclusion, fuzzy logic tools have the ability for find consistent patterns or relationship between factors of complex process and generate understandable knowledge in an explicitly format (linguistic rules) [60, 61].

Fuzzy logic can be combined with ANNs forming neurofuzzy logic. This is a hybrid system that combines the adaptive learning capabilities from ANNs with the flexibility of representation of fuzzy logic [55].

5. Applications of artificial intelligence tools in seed germination and dormancy

AI is a novel technology in plant biology, and only a very scarce literature has published applying AI tools to germination and dormancy. However, these works are very interesting since they suggest that AI tools may predict and optimize germination and dormancy processes.

As previously discussed, a representative study of seed germination requires large experimental designs to include the effect of multiple factors (**Figure 2**). Therefore, the experimental design implies many treatments, replicates and, a huge number of seeds. However, not always is available a suitable size sample of seeds to draw clear conclusions. Under those circumstances, AI tools are an excellent alternative to conventional statistical methods. Advantageously, neurofuzzy logic technology allows working with not well-defined design spaces (reduced number of treatments) and different kind of data at the same time [62, 63].

The first papers published using AI in germination [64, 65], were devoted to predict field seed weed emergence, since is essential for minimizing economic losses and improve crop yield and management. Then, they really focus more in weed management and control than in elucidate the factors involved in seed weed germination. In those works, germination of the weed *Avena fatua*, was predicted with more accuracy with ANN models than with nonlinear regression analysis [64, 65]. In addition, those models were improved including some dormancy parameters such as after-ripening and implemented a genetic algorithm to optimize them. In this optimization process the mean square error between their experimental and training data were minimized. Therefore, they allow to obtain more parsimonious models and with better predictive capacity [66, 67].

Most germination research was carried out on seeds with commercial interest, however, many wild species, that never have been cultivated by humans, have deep dormancy and present seeds with underdeveloped embryos and with physiological dormancy. Moreover, some of those plants with poor germination are classified as vulnerable and endangered plants [68]. This is the case of *Eryngium viviparum* a threatened plant belonging to Apiaceae. This family has many species are well-known by a nonuniformity germination due to MD and MPD seeds. Recently [68], the hybrid neurofuzzy logic tool was used to decipher the relationship among several dormancy-breaking and germination factors (inputs) and several parameters (outputs), such as germination rate and embryo growth (E:S ratio). Neurofuzzy models allowed to found the most critical factors involved in the seed responses. In addition, IF-THEN rules pointed out the interaction of those factors to increase or promote the germination rate and the E:S ratio. The model revealed that the best germination rates were obtained with the combination of

1 mg L^{-1} GA_3 (gibberellic acid) and high (24°C) incubation temperature and the combination of long incubation (20 weeks) and short warm (25°C) stratification periods (4 weeks) [68].

More recently, also neurofuzzy logic was used successfully in order to discover the critical factors that break dormancy and increase the germination rate in several kiwifruit cultivars, then allow to describe the best conditions for kiwifruit seeds germination [69]. The next factors were investigated: a) the effect of stratification time and type on dormancy-breaking and the effect of thermo-photoperiod on germination. The results obtained demonstrated that neurofuzzy logic models greatly facilitate the data analysis and pointed out the critical role of cold-stratification time (long periods at 4°C) and stratification treatment (using gibberellic acid) on kiwifruit seed germination. In conclusion, neurofuzzy was able to model with high accuracy and predictability, to obtain a set of rules very useful for understand the cause-effect among the studied factors and dormancy-breaking and germination [68, 69].

6. Conclusions and future perspectives

Seed germination is a very complex bioprocess, dependent on many interacting factors. This kind of processes are not fully understood due to experimental limitations (low number of factors studied simultaneously or poorly designed experiments), which do not allow to study simultaneously all interactions among the factors involved. Currently, due to the emergence of computer-based technologies such as AI tools, those bottlenecks can be avoided. Artificial intelligence tools provide useful algorithms for studying complex processes, big datasets, being a quite novel technology in seed science.

In recent papers, ANNs combined with fuzzy logic had allowed to predict the germination of weed species, in a much easier way than traditional methods as statistical regressions. In addition, the use of ANNs combined with genetic algorithms allows to build up computer models to optimize, with high accuracy, the germination of these weeds, and hence, decreasing the economic losses in crop production. Moreover, the hybrid AI tool, neurofuzzy, demonstrated to be a successful technology to decipher the most critical factors and their interactions for increased germination and reduced the dormancy impact in some species.

In a near future, it seems that AI tools would be essential and very useful tools in germination studies, for the selection of most critical factors, with good accurateness in decipher the interaction between environment and physiological factors on dormancy and germination.

Acknowledgements

This research was supported by TREMEDAL-Inland wetlands of Northern Iberian Peninsula: management and restoration of mires and wet environments European Union (LIFE11 NAT/ES/000707, 2012-2015). This work was also funded by Xunta de Galicia, Spain (CITACA Strategic Partnership, Reference: ED431E 2018/07) and "Red de Uso Sostenible de Recursos y Residuos" (ED431D 2017/18).

Conflict of interest

The authors declare no conflict of interest.

Author details

Manuel Ayuso[1], Mariana Landín[2], Pedro Pablo Gallego[1*] and Mª Esther Barreal[1]

1 Applied Plant and Soil Biology, Faculty of Biology, University of Vigo, Vigo, Spain

2 Pharmacy and Pharmaceutical Technology Dpt., Faculty of Pharmacy, University of Santiago de Compostela, Santiago de Compostela, Spain

*Address all correspondence to: pgallego@uvigo.es

References

[1] Bewley J. Seeds of hope; seeds of conflict. In: Nicolas G, Bradford K, Come D, Pritchard H, editors. Biology of Seeds: Recent Research Advances. 1st ed. Wallingford, UK: CAB International; 2003. pp. 1-4. DOI: 10.1017/S001447970425205X

[2] Bewley JD, Bradford KJ, Hilhorst HWM, Nonogaki H. Seeds: Physiology of Development, Germination and Dormancy. 3rd ed. New York: Springer-Verlag; 2013. 392 p. DOI: 10.1007/978-1-4614-4693-4

[3] Baskin JM, Baskin CC. A classification system for seed dormancy. Seed Science Research. 2004;**14**:1-16. DOI: 10.1079/SSR2003150

[4] Baskin CC, Baskin JM. Seeds: Ecology, Biogeography, and Evolution of Dormancy and Germination. 2nd ed. San Diego: Elsevier Science; 2014. 1601 p. DOI: 10.1016/B978-0-12-080260-9.X5000-3

[5] Finch-Savage WE, Leubner-Metzger G. Seed dormancy and the control of germination. New Phytologist. 2006;**171**:501-523. DOI: 10.1111/j.1469-8137.2006.01787.x

[6] Struik PC, Yin X, de Visser P. Complex quality traits: Now time to model. Trends in Plant Science. 2005;**10**:513-516. DOI: 10.1016/J.TPLANTS.2005.09.005

[7] Gallego PP, Gago J, Landin M. Artificial neural networks technology to model and predict plant biology process. In: Suzuki K, editor. Artificial Neural Networks - Methodological Advances and Biomedical Applications. 1st ed. Rijeka, Croatia: IntechOpen; 2011. pp. 197-217. DOI: 10.5772/14945

[8] Hudson DL, Cohen ME. Neural Networks and Artificial Intelligence for Biomedical Engineering. 1st ed. New York: Institute of Electrical and Electronics Engineers; 1999. 306 p. DOI: 10.1109/9780470545355

[9] Huang Y. Advances in artificial neural networks – Methodological development and application. Algorithms. 2009;**2**:973-1007. DOI: 10.3390/algor2030973

[10] Cartwright H. Using Artificial Intelligence in Chemistry and Biology. 1st ed. Boca Raton: CRC Press; 2008 360 p

[11] Gago J, Landín M, Gallego P. Strengths of artificial neural networks in modeling complex plant processes. Plant Signaling & Behavior. 2010;**5**:743-745. DOI: 10.4161/psb.5.6.11702

[12] Gago J, Martínez-Núñez L, Landín M, Gallego P. Artificial neural networks as an alternative to the traditional statistical methodology in plant research. Journal of Plant Physiology. 2010;**167**:23-27. DOI: 10.1016/j.jplph.2009.07.007

[13] Finkelstein R, Reeves W, Ariizumi T, Steber C. Molecular aspects of seed dormancy. Annual Review of Plant Biology. 2008;**59**:387-415. DOI: 10.1146/annurev.arplant.59.032607.092740

[14] Nikolaeva M. Factors controlling the seed dormancy pattern. In: Khan A, editor. The Physiology & Biochemistry of Seed Dormancy and Germination. Amsterdam: North-Holland Publ. Co; 1977. pp. 51-74

[15] Bewley JD. Seed germination and dormancy. The Plant Cell. 1997;**9**:1055-1066. DOI: 10.1105/tpc.9.7.1055

[16] Lefebvre V, North H, Frey A, Sotta B, Seo M, Okamoto M, et al. Functional analysis of Arabidopsis *NCED6* and *NCED9* genes indicates that ABA synthesized in the endosperm is involved in the

induction of seed dormancy. The Plant Journal. 2006;**45**:309-319. DOI: 10.1111/j.1365-313X.2005.02622.x

[17] Cadman CSC, Toorop PE, Hilhorst HWM, Finch-Savage WE. Gene expression profiles of Arabidopsis cvi seeds during dormancy cycling indicate a common underlying dormancy control mechanism. The Plant Journal. 2006;**46**:805-822. DOI: 10.1111/j.1365-313X.2006.02738.x

[18] Fenner M, Thompson K. The Ecology of Seeds. 1st ed. Cambridge: Cambridge University Press; 2005. 250 p. DOI: 10.1017/CBO9780511614101

[19] Gutterman Y. Maternal effects on seeds during development. In: Fenner M, editor. Seeds : The Ecology of Regeneration in Plant Communities. Vol. 2. Wallingford,UK: CABI Pub; 2000. p. 410

[20] Nambara E, Marion-Poll A. ABA action and interactions in seeds. Trends in Plant Science. 2003;**8**:213-217. DOI: 10.1016/S1360-1385(03)00060-8

[21] Raz V, Bergervoet JH, Koornneef M. Sequential steps for developmental arrest in Arabidopsis seeds. Development. 2001;**128**:243-252

[22] Finkelstein RR, Gampala SSL, Rock CD. Abscisic acid signaling in seeds and seedlings. The Plant Cell. 2002;**14**(Suppl):S15-S45. DOI: 10.1105/tpc.010441

[23] Robichaud CS, Wong J, Sussex IM. Control of in vitro growth of viviparous embryo mutants of maize by abscisic acid. Developmental Genetics. 1979;**1**:325-330. DOI: 10.1002/dvg.1020010405

[24] Nambara E, Hayama R, Tsuchiya Y, Nishimura M, Kawaide H, Kamiya Y, et al. The role of *ABI3* and *FUS3* loci in *Arabidopsis thaliana* on phase transition from late embryo development to germination. Developmental Biology. 2000;**220**:412-423. DOI: 10.1006/DBIO.2000.9632

[25] Hilhorst HWM. A critical update on seed dormancy. I. Primary dormancy. Seed Science Research. 1995;**5**:61-73. DOI: 10.1017/S0960258500002634

[26] Kucera B, Cohn MA, Leubner-Metzger G. Plant hormone interactions during seed dormancy release and germination. Seed Science Research. 2005;**15**:281-307. DOI: 10.1079/SSR2005218

[27] Grappin P, Bouinot D, Sotta B, Miginiac E, Jullien M. Control of seed dormancy in *Nicotiana plumbaginifolia*: Post-imbibition abscisic acid synthesis imposes dormancy maintenance. Planta. 2000;**210**:279-285. DOI: 10.1007/PL00008135

[28] Le Page-Degivry MT, Garello G. In situ abscisic acid synthesis: A requirement for induction of embryo dormancy in *Helianthus annuus*. Plant Physiology. 1992;**98**:1386-1390. DOI: 10.1104/pp.98.4.1386

[29] Wang M, Heimovaara-Dijkstra S, Van Duijn B. Modulation of germination of embryos isolated from dormant and nondormant barley grains by manipulation of endogenous abscisic acid. Planta. 1995;**195**:586-592. DOI: 10.1007/BF00195719

[30] Liu X, Zhang H, Zhao Y, Feng Z, Li Q, Yang H-Q, et al. Auxin controls seed dormancy through stimulation of abscisic acid signaling by inducing ARF-mediated *ABI3* activation in *Arabidopsis*. Proceedings of the National Academy of Sciences of the United States of America. 2013;**110**:15485-15490. DOI: 10.1073/pnas.1304651110

[31] Shu K, Liu X, Xie Q, He Z. Two faces of one seed: Hormonal regulation of dormancy and germination.

Molecular Plant. 2016;**9**:34-45. DOI: 10.1016/J.MOLP.2015.08.010

[32] Barrero JM, Jacobsen JV, Talbot MJ, White RG, Swain SM, Garvin DF, et al. Grain dormancy and light quality effects on germination in the model grass Brachypodium distachyon. The New Phytologist. 2012;**193**:376-386. DOI: 10.1111/j.1469-8137.2011.03938.x

[33] Millar AA, Jacobsen JV, Ross JJ, Helliwell CA, Poole AT, Scofield G, et al. Seed dormancy and ABA metabolism in *Arabidopsis* and barley: The role of ABA 8′-hydroxylase. The Plant Journal. 2006;**45**:942-954. DOI: 10.1111/j.1365-313X.2006.02659.x

[34] Gubler F, Hughes T, Waterhouse P, Jacobsen J. Regulation of dormancy in barley by blue light and after-ripening: Effects on abscisic acid and gibberellin metabolism. Plant Physiology. 2008;**84**:61-66. DOI: 10.1104/pp.84.1.61

[35] Lee S, Cheng H, King KE, Wang W, He Y, Hussain A, et al. Gibberellin regulates *Arabidopsis* seed germination via RGL2, a GAI/RGA-like gene whose expression is up-regulated following imbibition. Genes and Development. 2002;**16**:646-658. DOI: 10.1101/gad.969002

[36] Yamauchi Y, Takeda-Kamiya N, Hanada A, Ogawa M, Kuwahara A, Seo M, et al. Contribution of gibberellin deactivation by AtGA2ox2 to the suppression of germination of dark-imbibed *Arabidopsis thaliana* seeds. Plant and Cell Physiology. 2007;**48**:555-561. DOI: 10.1093/pcp/pcm023

[37] Jacobsen SE, Olszewski NE. Mutations at the SPINDLY locus of *Arabidopsis* alter gibberellin signal transduction. The Plant Cell. 1993;**5**:887-896. DOI: 10.1105/tpc.5.8.887

[38] Bethke PC, Libourel IGL, Aoyama N, Chung Y-Y, Still DW, Jones RL. The Arabidopsis aleurone layer responds to nitric oxide, gibberellin, and abscisic acid and is sufficient and necessary for seed dormancy. Plant Physiology. 2007;**143**:1173-1188. DOI: 10.1104/pp.106.093435

[39] Holman TJ, Jones PD, Russell L, Medhurst A, Tomás SÚ, Talloji P, et al. The N-end rule pathway promotes seed germination and establishment through removal of ABA sensitivity in *Arabidopsis*. Proceedings of the National Academy of Sciences. 2009;**106**:4549-4554. DOI: 10.1073/PNAS.0810280106

[40] Graeber K, Nakabayashi K, Miatton E, Leubner-Metzger G, Soppe W. Molecular mechanisms of seed dormancy. Plant, Cell & Environment. 2012;**35**:1769-1786. DOI: 10.1111/j.1365-3040.2012.02542.x

[41] Footitt S, Douterelo-Soler I, Clay H, Finch-Savage WE. Dormancy cycling in Arabidopsis seeds is controlled by seasonally distinct hormone-signaling pathways. Proceedings of the National Academy of Sciences. 2011;**108**:20236-20241. DOI: 10.1073/PNAS.1116325108

[42] Allen PS, Benech-Arnold RL, Batlla D, Bradford KJ. Modeling of seed dormancy. In: Bradford KJ, Nonogaki H, editors. Seed Development, Dormancy and Germination. 1st ed. Chichester, UK: John Wiley & Sons, Ltd; 2007. pp. 72-112

[43] Probert RJ. The role of temperature in the regulation of seed dormancy and germination. In: M Fenner, editor. Seeds: The Ecology of Regeneration in Plant Communities. Wallingford: CABI; 2000. pp. 261-292

[44] Commander LE, Merritt DJ, Rokich DP, Dixon KW. The role of after-ripening in promoting germination of arid zone seeds: A study on six Australian species. Botanical Journal of the Linnean Society. 2009;**161**:411-421. DOI: 10.1111/j.1095-8339.2009.01009.x

[45] Yamauchi Y, Ogawa M, Kuwahara A, Hanada A, Kamiya Y, Yamaguchi S. Activation of gibberellin biosynthesis and response pathways by low temperature during imbibition of *Arabidopsis thaliana* seeds. The Plant Cell. 2004;**16**:367-378. DOI: 10.1105/tpc.018143

[46] Chiwocha SDS, Cutler AJ, Abrams SR, Ambrose SJ, Yang J, Ross ARS, et al. The *etr1-2* mutation in *Arabidopsis thaliana* affects the abscisic acid, auxin, cytokinin and gibberellin metabolic pathways during maintenance of seed dormancy, moist-chilling and germination. The Plant Journal. 2005;**42**:35-48. DOI: 10.1111/j.1365-313X.2005.02359.x

[47] Otroshy M, Zamani A, Khodambash M, Ebrahimi M, Struik PC. Effect of exogenous hormones and chilling on dormancy breaking of seeds of asafoetida (*Ferula assafoetida* L.). Research Journal of Seed Science. 2009;**2**:9-15. DOI: 10.3923/rjss.2009.9.15

[48] Subhashini Devi P, Satyanarayana B, Arundhati A, Raghava Rao T. Effect of storage temperature and dormancy-breaking treatments on seed germination, moisture content and seed vigor in gum karaya (*Sterculia urens* Roxb.). Forest Science and Technology. 2012;**8**:11-15. DOI: 10.1080/21580103.2012.658235

[49] Schütz W. Ecology of seed dormancy and germination in sedges (Carex). Perspectives in Plant Ecology, Evolution and Systematics. 2000;**3**:67-89. DOI: 10.1078/1433-8319-00005

[50] Baskin CC, Baskin JM. Germination ecophysiology of herbaceous plant species in a temperate region. American Journal of Botany. 1988;**75**:286-305. DOI: 10.1002/j.1537-2197.1988.tb13441.x

[51] Li W, Liu X, Ajmal Khan M, Yamaguchi S. The effect of plant growth regulators, nitric oxide, nitrate, nitrite and light on the germination of dimorphic seeds of *Suaeda salsa* under saline conditions. Journal of Plant Research. 2005;**118**:207-214. DOI: 10.1007/s10265-005-0212-8

[52] Pérez-Piñeiro P, Gago J, Landín M, Gallego PP. Agrobacterium-mediated transformation of wheat: General overview and new approaches to model and identify the key factors involved. In: Ozden Y, editor. Transgenic Plants: Advances and Limitations. Rijeka, Croatia: InTechOpen; 2012. pp. 3-26

[53] Yuan JS, Galbraith DW, Dai SY, Griffin P, Stewart CN. Plant systems biology comes of age. Trends in Plant Science. 2008;**13**:165-171. DOI: 10.1016/J.TPLANTS.2008.02.003

[54] Gago J, Martínez-Núñez L, Landín M, Flexas J, Gallego P. Modeling the effects of light and sucrose on In vitro propagated plants: A multiscale system analysis using artificial intelligence technology. PLoS One. 2014;**9**:e85989. DOI: 10.1371/journal.pone.0085989

[55] Shao Q, Rowe RC, York P. Comparison of neurofuzzy logic and decision trees in discovering knowledge from experimental data of an immediate release tablet formulation. European Journal of Pharmaceutical Sciences. 2007;**31**:129-136. DOI: 10.1016/J.EJPS.2007.03.003

[56] Jiménez D, Pérez-Uribe A, Satizábal H, Barreto M, Van Damme P, Tomassini M. A Survey of Artificial Neural Network-Based Modeling in Agroecology. Soft Computing Applications in Industry. 1st ed. Berlin, Heidelberg: Springer Berlin Heidelberg; 2008. pp. 247-269. DOI: 10.1007/978-3-540-77465-5_13

[57] Hilbert DW, Ostendorf B. The utility of artificial neural networks for modelling the distribution of vegetation in past, present and future climates.

Ecological Modelling. 2001;**146**:311-327. DOI: 10.1016/S0304-3800(01)00323-4

[58] Plumb AP, Rowe RC, York P, Brown M. Optimisation of the predictive ability of artificial neural network (ANN) models: A comparison of three ANN programs and four classes of training algorithm. European Journal of Pharmaceutical Sciences. 2005;**25**:395-405. DOI: 10.1016/J.EJPS.2005.04.010

[59] Osama K, Mishra BN, Somvanshi P. Machine learning techniques in plant biology. In: Barh D, Khan M, Davies E, editors. PlantOmics: The Omics of Plant Science. 1st ed. New Delhi: Springer India; 2015. pp. 731-754

[60] Landín M, Rowe RC, York P. Advantages of neurofuzzy logic against conventional experimental design and statistical analysis in studying and developing direct compression formulations. European Journal of Pharmaceutical Sciences. 2009;**38**:325-331. DOI: 10.1016/J.EJPS.2009.08.004

[61] Gago J, Landín M, Gallego PP. A neurofuzzy logic approach for modeling plant processes: A practical case of in vitro direct rooting and acclimatization of Vitis vinifera L. Plant Science. 2010;**179**:241-249. DOI: 10.1016/J.PLANTSCI.2010.05.009

[62] Nezami-Alanagh E, Garoosi G-A, Maleki S, Landín M, Gallego P. Predicting optimal in vitro culture medium for Pistacia vera micropropagation using neural networks models. Plant Cell, Tissue and Organ Culture (PCTOC). 2017;**129**:19-33. DOI: 10.1007/s11240-016-1152-9

[63] Nezami-Alanagh E, Garoosi G, Haddad R, Maleki S, Landín M, Gallego P. Design of tissue culture media for efficient Prunus rootstock micropropagation using artificial intelligence models. Plant Cell Tissue and Organ Culture. 2014;**117**:349-359. DOI: 10.1007/s11240-014-0444-1

[64] Chantre GR, Blanco AM, Lodovichi MV, Bandoni AJ, Sabbatini MR, López RL, et al. Modeling Avena fatua seedling emergence dynamics: An artificial neural network approach. Computers and Electronics in Agriculture. 2012;**88**:95-102. DOI: 10.1016/J.COMPAG.2012.07.005

[65] Blanco AM, Chantre GR, Lodovichi MV, Bandoni JA, López RL, Vigna MR, et al. Modeling seed dormancy release and germination for predicting *Avena fatua* L. field emergence: A genetic algorithm approach. Ecological Modelling. 2014;**272**:293-300. DOI: 10.1016/J.ECOLMODEL.2013.10.013

[66] Chantre G, Blanco A, Forcella F, Van Acker R, Sabbatini M, Gonzalez-Andujar J. A comparative study between non-linear regression and artificial neural network approaches for modelling wild oat (*Avena fatua*) field emergence. The Journal of Agricultural Science. 2014;**152**:254-262. DOI: 10.1017/S0021859612001098

[67] Chantre GR, Vigna MR, Renzi JP, Blanco AM. A flexible and practical approach for real-time weed emergence prediction based on artificial neural networks. Biosystems Engineering. 2018;**170**:51-60. DOI: 10.1016/J.BIOSYSTEMSENG.2018.03.014

[68] Ayuso M, Ramil-Rego P, Landin M, Gallego PP, Barreal ME. Computer-assisted recovery of threatened plants: Keys for breaking seed dormancy of *Eryngium viviparum*. Frontiers in Plant Science. 2017;**8**:2092. DOI: 10.3389/fpls.2017.02092

[69] González-Puelles JE, Landín M, Gallego PP, Barreal ME. Deciphering kiwifruit seed germination using neural network tools. Acta Horticulturae. 2018;**1218**: 359-366. DOI: 10.17660/ActaHortic.2018.1218.50

Section 2

Germination

Chapter 4

Programmed Cell Death in Seeds: An Adaptive Mechanism Required for Life

Angel J. Matilla

Abstract

The regeneration of the mother plant through germinative process is the main reason that evolutionarily justifies the existence of a viable seed. Current knowledge indicates that the control of germination is a sophisticated process mainly controlled by hormones and reactive oxygen species (ROS), among other endogenous factors. One of the events that directly participate in the germination is the degradation of storage proteins (SPs). Thus, vacuolar processing enzymes (VPEs) contribute to SPs' degradation and mobilization due to direct proteolysis or through the activation of other peptidases. In parallel, the relationship between VPEs and programmed cell death (PCD) is beyond doubt. As an alternative to VPEs, the formation of vesicles called ricinosomes containing papain-like Cys-proteases (PLCPs) and located in the reserve tissues of some germinating seeds also collaborates to protein degradation. Finally, there are increasing evidences linking nucleases to PCD in different tissues of seed. However, its state of the art is still little developed. Together, this current overview illustrates a part of the complexity of PCD in seeds, a puzzle far from being solved.

Keywords: Cys-endoproteases, endosperm, lytic vacuoles, nucleases, papain-like Cys-proteases, ricinosomes, seeds, seed storage proteins, vacuolar processing enzymes

1. Introduction

The life cycle of organisms requires targeted cell types to be removed in a predictable and genetically organized way. This process of cellular suicide, named programmed cell death (PCD), occurs from embryogenesis to senescence and is an essential part of development and cell homeostasis of any multicellular organism [1–3]. Thus, PCD has been observed from the onset of zygotic embryogenesis until the germinative process ends [4–6]. The mechanism through which specific cells are targeted for PCD without affecting neighboring cells has not yet been resolved. Notable cellular compartments (i.e., mitochondria, chloroplasts, Golgi complex, endoplasmic reticulum (ER), and vacuoles) have been shown to be involved in PCD [7]. Plant PCD exhibits several hallmarks: (i) DNA laddering and strong chromatin condensation [8]; (ii) sometimes, release of cytochrome-c from the mitochondria to the cytosol, and its subsequent degradation, which is dependent on reactive oxygen species (ROS) and caspase-like activity [9]; (iii) generation of autophagic vacuoles due to the absence of an active phagocytosis system [10, 11];

IntechOpen

(iv) degradation of organelles such as the plastidome, mitochondria, and peroxisomes [11]; (v) extensive vacuolation (i.e., appearance of a large vacuole) [12]; (vi) sometimes, development of ricinosomes concomitantly with the progression of nuclear DNA fragmentation [13, 14]; and (vii) contribution of nucleases and ROS [15, 16]. At the end of PCD, the cell is completely digested, and the remaining protoplast is surrounded by the cell wall (CW), which finally becomes disorganized and disintegrates in a coordinated and regulated way [17]. Because plants have CWs, they have developed their own PCD process, thus not requiring the apoptotic regulators and phagocytic processes present in animal cells.At the cellular level, plant PCD can be non-autolytic or autolytic (i.e., formation of large lytic vacuoles and rapid clearance of cytoplasm due to tonoplast rupture and the release of active hydrolases) [18]. Thus, developmental PCD (dPCD) is autolytic and is critical for many vegetative and reproductive processes [2, 19, 20]. However, environmental PCD (ePCD) is non-autolytic and is involved in responses to biotic and abiotic stresses. In this latter form of PCD is involved the hypersensitive response (HR), which prevents the growth and spread of pathogens into healthy tissues [21–23]. Recently, it has been suggested that dPCD and ePCD are characterized by separate regulatory pathways. In fact, a conserved core of transcriptionally controlled dPCD-associated genes has been defined [24]. Because plants and animals have different molecular mechanisms for PCD, an evolutionary parallelism of PCD pathways in plants and animals has been postulated [25].

The involvement of PCD has been described in various plant life processes, including the emptying of xylem tracheary elements [26], aerenchyma formation [25, 27], and dynamic turnover of the root cap [28]. In addition, PCD is an integral part of the seed development and germination (i.e., dPCD), during which cells of the integuments, nucellus, suspensor, and endosperm face death [5, 6]. The following text presents an update on the substantial progress that has been made to our understanding of PCD through the life of the seed, an entity that represents the dispersal unit of the spermatophytes securing their survival and perpetuation. The role of papain-type KDEL-Cys-endoproteases (PLCPs), vacuolar processing enzymes (VPEs) and nucleases, is carefully reviewed.

2. The role of plant-specific KDEL-Cys-endopeptidases in seed development and germination

2.1 Ricinosomes

Cys-endopeptidases (Cys-EPs) are the most abundant group of proteases responsible for degradation and the mobilization of storage proteins (SPs), being the SPs of seeds the most affected [29]. Cys-EP is a member of a unique group of papain-type Cys-EPs found specifically during senescence. The ER-derived vesicles (e.g., protein bodies, glyoxysomes, and ricinosomes) accumulate in seeds, among other compounds, specific SPs, (e.g., prolamin and zein) and KDEL-tailed and papain-type proteases [30, 31]. The SPs' accumulation process is mediated by ER chaperones such as the luminal binding protein (BiP) and protein disulfide isomerase (PDI). Interestingly, BiP can function either as a negative or a positive modulator of PCD events and also participate in innate immunity. Besides, in the seeds of castor bean, the immature 11S globulin was aggregated and then packaged in vesicles from ER [32]. That is, the ER-derived vesicles are thought to function as repositories of specific proteins until they are required for the cellular metabolism.

The ricinosomes (**Figure 1**) are spherical plant-specific organelles that have been firstly documented in senescing germinating endosperms of castor bean

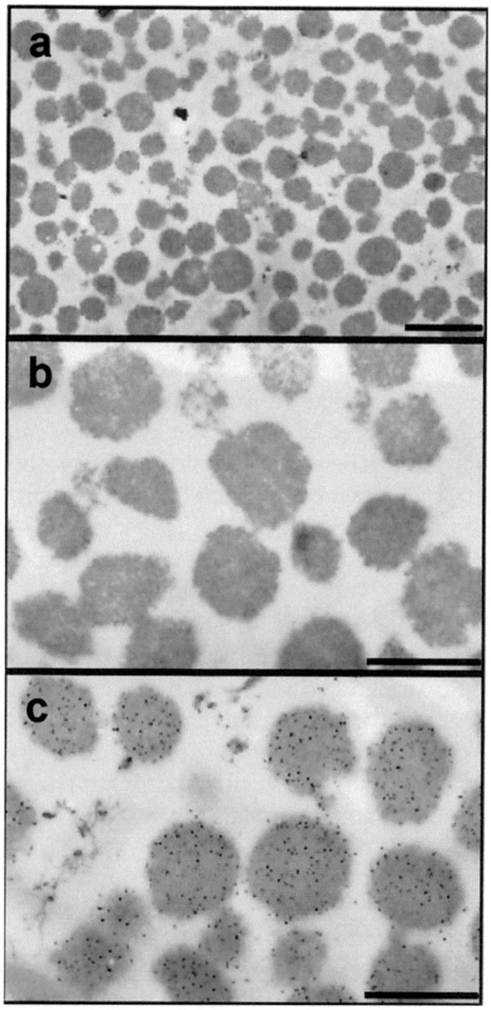

Figure 1.
Ultrastructure of ricinosomes purified from 5-day-old castor bean endosperm and immunolocalization of their marker enzyme Cys-EP. Electron micrographs (a, ×4400; b, ×12,000) and immunocytochemistry by using α-CysEP (c, ×12,000). Scale bar: a = 1.0 μm; b and c = 0.5 μm. From Schmid et al. [30] with permission of PNAS (USA).

[30, 33–35]. They are present prior to the appearance of other subcellular changes related to PCD and appear at the beginning of PCD and deliver large amounts of papain-type Cys-EPs in the final stages of cellular disintegration [13]. The ricinosomes contain large quantities of a 45-kDa pro-Cys-EP with a C-terminal KDEL (ER retention signal), and they are specifically for plant PCD [30, 36]. The ricinosomes are surrounded by a single ribosome-studded membrane and are directly sorted toward vacuoles through a Golgi-independent pathway to get involved in the PCD. These vesicles bud off from the ER in senescing tissues concomitantly with the progress of nuclear DNA fragmentation and have Cys-EPs as marker enzymes [37, 38]. KDEL-Cys-EPs are synthesized as inactive or weakly active pre-proenzyme which usually include a KDEL and an auto-inhibitory pro-domain that is cotranslationally transferred into the ER and then stored in ricinosomes because the pro-domain prevents premature activation of the protease [39]. Upon cytosolic acidification due to the LV collapse, the KDEL-Cys-EPs autocatalytic activation occurs [40]. This activation has been confirmed by in vitro acidification experiments of isolated ricinosomes and implies the cleavage of the N-terminal pro-peptide and the C-terminal KDEL motif. The presence of mature Cys-EP is consistent with the loss of tonoplast integrity. The mature and enzymatically active KDEL-Cys-EPs exhibit unusual broad substrate specificity (**Figure 2A**). This characteristic is due to the fact that the active site accepts a wide variety of amino acids, including proline and glycosylated hydroxyproline (e.g., extensions) from the glycoproteins of the CW [41]. When ricinosomes disintegrate and release their content into the cytoplasm, the cells that contain them are going to die [13, 14, 42]. More specifically, these ER vesicles are present prior to the appearance of other subcellular changes related to vacuolar cell death, one of the two classes of PCD previously defined [13, 35, 42, 43]. Interestingly, the ricinosomes, but not the enzymes within them, have also been implicated in the PCD of *Solanum lycopersicum* [35]. Likewise, anther dehiscence in tomato has also been linked to dPCD, and accumulation of ricinosome-like vesicles and the dPCD-associated SlCys-EP has been observed in the dehiscence zones of tomato anthers along with nuclear condensation and cytoplasmic retraction [13]. In year 2014, the first evidence for the existence of ricinosomes in *Arabidopsis* has been documented [44].

2.2 Involvement of papain-like KDEL-Cys-EPs in seed life

Papain-like Cys-EPs (PLCPs; often called cathepsins in animals) are essential and central hubs of plant immunity, germination, development, and senescence [45, 46]. Thus, when activated, PLCPs induce a broad spectrum of defense responses, including PCD [46]. On the other hand, PLCPs constitute one of the most abundant groups of the proteases responsible for the degradation and mobilization of SPs in seeds [47]. Their role during germination has been reported in a wide range of both monocot and dicot plants [48]. PLCPs in plants are divided

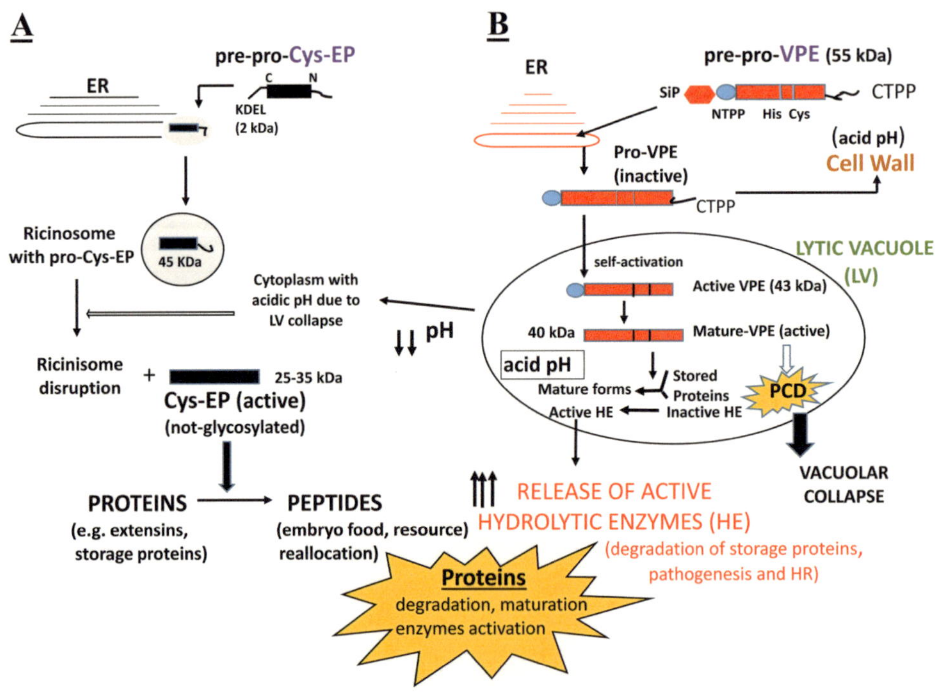

Figure 2.
Maturation, activation, and involvement of papain-type KDEL-Cys-EPs (A) and γVPE (B) in plant PCD. Endoplasmic reticulum (ER), N-terminal pro-peptide (NTPP), self-inhibitory C-terminal pro-peptide (CTPP), signal peptide (SiP), and storage protein (SP) (see text for more details).

into nine subfamilies. Thus, 32, 41, and 45 PLCPs' members have been identified in *Arabidopsis*, barley, and rice, respectively [49]. PLCPs have no structural relationship to the caspases, and its natural competitive and reversible inhibitors are the phytocystatins which are evolutionarily well conserved [50]. Recent results support the bifunctional ability of carboxy-extended phytocystatins in regulating legumain proteases via its carboxy-extended domain and PLCPs by its amino-terminal domain [51]. The activities of phytocystatins and PLCPs need to maintain a relatively balanced level to ensure the normal seed germination [29].

KDEL-tailed Cys-protease SH-EP is the first Cys-EP found to have a KDEL tail in spite of the fact that the protease localizes in the protein storage vacuoles [52]. KDEL-tailed protease-accumulating vesicles in germinating mung bean (*Vigna mungo*) cotyledons are similar to ricinosomes in that they accumulate the KDEL-tailed cysteine protease SH-EP [53, 54]. During the seeds' life, the ricinosomes accumulate PLCPs for the degradation of seed storage materials in both cotyledons and endosperm [30, 53]. Upon cell death, the content of ricinosomes (i.e., PLCPs) is released into the cell corpse where the proteinases are activated and proceed to degrade any remaining protein for the growing seedling in the case of nutritive seed tissues. Alternatively, PLCPs can also digest CW extensions in the final stage of PCD when the cell collapses and tissue breaks down [55]. Thus, the absence of ricinosomes during seed development (e.g., perisperm, integuments, chalaza, and pericarp) may be due to the fact that the CWs remain intact until germination, at which time these tissues are finally dismantled [56]. Interestingly, area micropylar of *Chenopodium quinoa* seeds does not have ricinosomes [6, 56]. KDEL-Cys-EPs are unique in digesting the extensions that form the basic scaffold for CW formation [55] (**Figure 2A**). So, KDEL-CPs like AtCEP1 are considered as late-acting proteases that digest CW proteins during the final stages of PCD and tissue remodeling after cellular disintegration [55, 57].

During seed germination, SPs are degraded to nourish the growing seedlings. This process is mainly triggered by PLCPs [29]. As a example, during both *Zea mays* and *Triticum aestivum* germination, the activity of Cys-EPs increases up to 90% of the total proteolytic activity. During barley seed germination, PLCPs were secreted from the scutellar and the aleurone layers to the endosperm to degrade the endosperm Sps [58]. Recently, the results of overexpression and silencing of *HvPap-1*, a gibberellin (GAs)-induced PLCP gen, indicated that PLCPs are important factors in mobilizing SPs to promote seed germination, and their expression and/or activity are regulated by GAs, ABA, and cystatins [49]. Ricinosomes and nuclear DNA are fragmented during PCD. In *Arabidopsis*, three KDEL-Cys-EPs called AtCEP1, AtCEP2, and AtCEP3 have been expressed in tissues undergoing PCD. Thus, the first gen is expressed in senescing ovules, the second in the vascular vessels, and the third in maturing siliques [55, 57–59]. Recently, AtCEP2 storing ricinosomes in *Arabidopsis* seedlings seems to be—like ER bodies—exclusively localized in epidermal cells [44]. The accumulation of KDEL-Cys-EPs and the appearance of ricinosomes may predict the occurrence of PCD during late seed development [37]. The ricinosomes containing pro-Cys-EP have been observed in anther tissues prior to PCD [13] and in the endosperm cells of imbibed tomato seeds (*Solanum lycopersicum*) where the reserve mobilization, Cys-EP accumulation and processing, is GA-induced [60]. Cereal aleurone PCD is controlled by phytohormones: the PCD promoting GAs and the antagonistically acting ABA [61]. The presence of ABA- and GA-responsive genes encoding proteases confirms their notable role in regulating the growth of cereal seeds [5, 62]. The endosperm in cereal seeds undergoes PCD during development, and, with the exception of the aleurone layer, is a dead tissue at maturity. In *Ricinus communis* the KDEL-Cys-EPs and ricinosomes were detected for the first time not only in the senescing endosperm of germinating seeds [30] but also in the nucellus of seeds during maturation [36, 63]. Ricinosomes with the pro-form of KDEL-Cys-EPs are also present in imbibed tomato seeds [60]. The presence of KDEL-Cys-EPs has been also demonstrated in (i) the hypogeous cotyledons of *Vicia sativa* [64]; (ii) the seed coat of *Phalaenopsis* [65]; (iii) the megagametophyte cells after germination of *Picea glauca* seeds [66]; (iv) the epigeous cotyledons of *Vigna mungo* [52]; (v) the senescing endosperm of germinating castor bean seeds [30, 67]; (vi) the nucellus in maturing castor bean seeds, where the endosperm expands at the expense of the nucellus cells [36]; (vii) the endosperm cells of imbibed tomato seeds [60]; and (viii) the germinating mung bean (*Vigna mungo*) cotyledons in that they accumulate the KDEL-tailed cysteine protease SH-EP [53]. Recently, an attractive PLCP protein called NbCP14 was characterized in *Nicotiana benthamiana*. This autocatalytically activated enzyme seems to be a Cath-H-like protease with great importance for the execution of PCD during plant development [68]. Previous to the NbCP14 identification, NtCP14 was also described in *Nicotiana tabacum* as a key component in triggering of PCD during the early stages of embryogenesis [69]. In the tobacco suspensor, PCD is antagonistically regulated by NtCP14 and its cystatin inhibitor NtCYS. Both silencing of NtCP14 and overexpression of NtCYS delay PCD [69].

3. Entailment of vacuolar processing enzymes with plant PCD

The cellular vacuoles execute essential functions for plant growth, development, and adaptation to biotic and abiotic stresses. In the absence of macrophages, the unwanted material for plant PCD is only degraded through vacuole-released hydrolytic enzymes, located in LVs (acidic pH) [70]. The formation of LVs involves the coalescence of protein storage vacuoles (PSVs, with a pH near neutrality), vacuolar lumen acidification, and intracellular material mobilization (i.e., cytoplasm engulfing). Only cells with high vacuolation resulted in PCD [12]. In brief, PSVs contain large amounts of defensive and SPs to be used during seed germination, while LVs contain hydrolytic enzymes [50]. Therefore, the degree of vacuolation can reflect the intensity of the PCD process. A clear example of vacuolization takes place in the aleurone cells during cereal seed germination [5, 12]. Mature cereal seeds consist mainly of dead cells, and only the embryo and aleurone layer are still alive. The PCD of aleurone cells is an essential process for the successful completion of post-germination storage mobilization, which is associated with the vacuole destruction [70]. Vacuolation and PCD of aleurone cells are initiated near the embryo and then gradually reach the distal area of the embryo [5, 12]. Similarly to the micro- or macroautophagy processes, the disruption of LVs and the concurrent release of various hydrolytic enzymes indicate that the PCD has been triggered [12]. The bursting of the tonoplast leads to a rapid cytoplasmic acidification and hydrolysis of the remaining cellular contents [71]. So much so, this vacuolar collapse needs to be rigorously organized to achieve PCD at a suitable timing [72]. Because the vacuolar collapse releases hydrolytic enzymes, the vacuole rupture is used as an indicator of PCD initiation [73]. Therefore, the tonoplast breakage is considered a point of no return during plant PCD [2]. In brief, the tonoplast rupture and vacuolar collapse are two important features of plant PCD. Finally, the plasmalemma integrity is maintained until the vacuole collapses [12, 74–76].

Plant PCD is accompanied by the upregulation of a heterogeneous group of vacuolar hydrolytic enzymes, being the vacuolar processing enzymes (VPEs), also called legumains, closely involved in its activation. VPEs originate from prokaryote pro-legumains. The VPEs have properties similar to animal caspases and fulfill relevant vacuolar functions in seeds [77]. Nevertheless, VPEs are directly involved in plant development and environmental stress responses [50]. Earlier studies in pumpkin seeds have demonstrated the identification of VPE as a vacuolar Cys-EP protein probably responsible for degradation of vacuolar SPs during germination [78]. Thus, the VPEs in monocots (i) are required for processing of glutelins that are the dominant seed SPs in rice [79], and (ii) they also process other seed SPs such as albumins, globulins, and ricins in storage vacuoles in seeds of pumpkin and castor bean [80]. However, VPE deficiency does not affect storage protein degradation in germinating seeds [81]. VPE was the first identified enzyme in plants with both caspase-like activity and activity against caspase-1-substrate [82]. Recent review contains the contributions of VPEs to plant PCD and its role in vacuole-mediated cell death [83]. Thus, the VPE4 expression pattern in the developing pericarp of *Nicotiana benthamiana* coincides with the profile of the caspase-1-like activity [84, 85]. Once vacuolar hydrolytic enzymes are activated, the proteolytic cascade leading PCD begins [70, 71]. However, although it is beyond question that VPE is an initiator of the vacuolar processing system, the mechanism by which VPE controls the vacuolar breakage and the execution of a variety of plant PCD is still unclear. In this regard, it was suggested that the disruption of the vacuole may be mediated by VPE in conjunction with protein kinases [86]. It has been also proposed that VPE and cathepsin-B (Cath-B), which have, respectively, caspase-1-like and caspase-3-like activity, may promote coalescence, accelerating the process of vacuolation and thus triggering vacuolar collapse during the PCD [87, 88]. However, no research yet has integrated the action of both VPE and Cath-B in the PCD pathway. VPE of *N. benthamiana* has been reported to mediate virus-induced HR by regulating tonoplast collapse [87, 89]. The PCD triggered by vacuolar collapse is unique to plants and has not been seen in animals (**Figure 2B**). As a result of this collapse and the liberation of active vacuolar hydrolases, the chromatin structure crumbled, the DNA is fragmented, and the plasma membrane disabled. Finally, the disintegration of the nuclear envelope starts [90], and the protoplast rapidly collapses and dies [5, 91].

Autophagy is a process known to mediate the degradation of residual proteins and aggregates of insoluble proteins and lipids and to remove damaged organelles. Likewise, autophagy is essential for vacuolation of cells undergoing developmental PCD and is activated by type-II metacaspases (McIIPa) [92]. Thus, during spruce embryogenesis McIIPa is transported from cytosol to the nucleus, where its presence is correlated to DNA fragmentation. These data reinforce that McIIPa is directly involved in a pathway which generates nuclear degradation, an event present in most programs of eukaryotic PCD. This McIIPa metacaspase can play a role on the cleavage of nuclear proteins [71, 93]. Besides metacaspases, VPE has been also described as another class of Cys-EPs involved besides in different types of PCD and also in development and immunity [70, 87, 94, 95]. The VPE contains a His-Cys catalytic dyad and cleaves a peptide bond at the C-terminal side involving an Asn residue, hence the name of asparaginyl endopeptidases [96]. VPEs are evolutionarily related to caspases and preferably localized in vacuoles (i.e., maximal activity at acidic pH) and are specific for plants [38, 70, 97]. Therefore, the plants might have evolved a VPE-mediated vacuolar system as a cellular suicide strategy. Plants encode at least four functional isoforms of VPEs, which are located in the vacuole ([77] and refs. therein). Plant VPEs are classified into vegetative, embryogenic, and seed-expressed types [98–100] (**Figure 3**). The genome of tomato has 14 VPE genes [95]. However, the genome of *Arabidopsis* has four VPE genes: α-VPE and β-VPE play a key role in the processing of SPs during seed maturation [80, 94], while γ-VPE and δ-VPE are expressed at early stage of seed development being involved in the formation of the inner integuments of the seed coat [101–103].

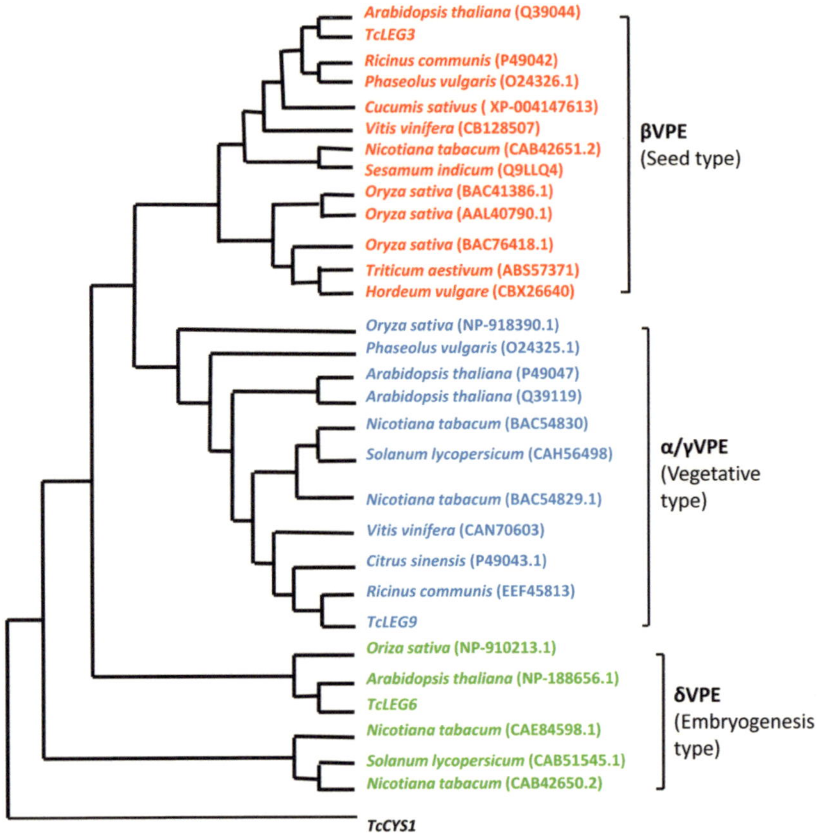

Figure 3.
Dendrogram of VPEs of several plant species. VPEs (access numbers are within parentheses) were separated into three groups: βVPE (seed type), α/γVPE (vegetative type), and δVPE (embryogenesis type). Signal peptides were excluded from the sequences. Adapted mainly from Nakaune et al. [103] and other recent publications.

Interestingly, in spite of delayed vacuolation, *Arabidopsis* γ-VPE mutants have a normal germination phenotype. This suggests that vacuolation does not trigger, but rather is a consequence of germination [104].

The δ-VPE was originated early during dicotyledonous diversification [103]. Regarding the maturation of γ-VPE in *A. thaliana* (**Figure 2B**), it is to know that the N-terminal signal peptide of VPE pre-protein precursor is cotranslationally removed in the ER to produce VPE pro-protein. The transfer of pro-protein precursor to the acidic vacuole causes the self-catalytic conversion into an intermediate isoform by removal of the C-terminal inhibitory pro-peptide. The subsequent removal of the N-terminal pro-peptide produces the mature γ-VPE [96]. In the case of the seed SPs, the VPEs' vacuolar maturation is of major importance, as it conditions the establishment of vigorous seedlings [105]. A quadruple-KO mutant with no detectable VPE activity strongly suggests that there are no other proteases with a similar activity in *Arabidopsis* [106]. When the VPE genes were knocked out, no characteristics belonging to cell death were observed [12]. VPE orthologs are widely distributed in land plants including mosses (e.g., *Physcomitrella patens*) and ferns (e.g., *Ceratopteris richardii*) [83]. In rice, five VPE (OsVPE) genes are found [83, 107]. Phylogenetic analyses and gene expression studies have demonstrated that OsVPE2 (OsaLeg2) and OsVPE3 (OsaLeg3) are involved in H_2O_2-induced PCD and in salt-stressed seeds, whereas OsaLeg1, OsaLeg4, and OsaLeg5 would act as vegetative-related legumains [83]. The barley (*Hordeum vulgare*) genome contains eight VPE genes (HvVPEs) which are differentially expressed during vegetative and reproductive development [98].

The first increase in a cascade of caspase-1-, caspase-3-, caspase-4-, caspase-6-, and caspase-8-like activities in the endosperm of *Hordeum vulgare* seeds may be related to PCD in the nucellus [84, 85, 108]. The importance of pericarp PCD for proper development of the endosperm has been recently described [84]. The increase in caspase-1-like activity may be acquired by HvVPE2a (called nucellain), HvVPE2b, and HvVPE2d proteases which is exclusively expressed in nucellus and nucellar projection. The expression patterns of the HvPhS2 and HvPhS3, which are exclusively active in the nucellar projection, coincide with the caspase-6-like activity profile in the early endosperm fraction indicating that HvPhS1 and HvPhS2 may be responsible for the caspase-6-like activity [84, 85]. Caspase-1-, caspase-3-, and caspase-6-like activities are also localized in the degenerating nucellus of *Sechium edule* [109]. In the degenerating nucellar tissue of castor bean, proteomic analyses identified multiple proteases and protease inhibitors [108]. The MADS-box transcription factor called MADS29 has been suggested to promote nucellar degeneration through the regulation of Cys-EP expression in rice and maize [61]. The α, β, γ, and δ-VPEs of *Arabidopsis* appear to share no direct one-to-one relationships of orthology with VPEs from gymnosperms. The VPE protein and its transcripts increase at the beginning of the HR reaction in the tobacco leaf, in which the cells showed typical PCD characteristics, and both the VPE inhibitor ESEN-CHO and the caspase-1-like activity inhibitor Ac-YVAD-CHO inhibit the appearance of PCD [98, 99]. These and other recent results [100] reaffirm that VPE is a protease with caspase-1-like activity in plants. VPE activation was started once the leaves of tobacco were infected with the TMV, leading to vacuole disruption and activation of PCD to prevent the proliferation of virus [70, 87]. Likewise, PCD during HR is critical for the removal of biotrophic pathogens, whose growth depends on the living host tissues [87, 99]. Together, VPE deficiency suppresses vacuolar collapse, leading to mycotoxin-induced cell death [83]. Interestingly, the results of Zhan's group using the NbVPE silencing suggest that VPE plays an important role in elicitor signaling in plants of *Nicotiana benthamiana* [89]. Finally, and based on the results to date, (i) transcriptome sequence information has permitted the identification of new VPE genes than having a cyclization function rather a protease function ([77] and refs. therein), and (ii) VPEs and other vacuolar enzymes once released from LV to cytosol through a barely known route promote a VPE-mediated vacuolar disruption and constitute a fundamental piece in the plant PCD puzzle whose organization is far from unravelling.

4. Nucleases: the next frontier for knowledge of PCD in plants

As shown throughout this review, the PCD process involves the selective removal of unwanted cells and the mobilization of cellular debris, including the products of DNA fragmentation, which is a known hallmark of PCD [6, 17, 93, 110, 111]. In plants, genomic DNA is actively degraded during dPCD (e.g., during seed coat formation [103] and barley pericarp development [112]). Although the enzymes directly involved in nuclear dismantling are unknown, there is increasing evidence linking proteases and nucleases to plant PCD [20]. What is known is that during PCD, different nucleases are induced, including a set of S1-type (Zn^{+2} dependent) endonucleases that are synthesized regardless of tissue type [113, 114]. While these nucleases are cytoplasmic and lack a canonical nuclear localization signal, upon induction of cell death, they become nuclear [115]. Thus, nucleases are tightly associated with different plant PCD processes, including HR [116], endosperm development aleurone cell death [4, 117], and xylogenesis [118]. It has been hypothesized that PCD-associated nucleases help to recycle DNA from dead cells by degrading it into smaller fragments so that it can be taken up to be reused by neighboring cells. In cereal seeds, the progression of endosperm PCD is accompanied by an increase in nuclease activity and the degradation of nuclear DNA at internucleosomal sites [4, 115]. PCD in the endosperm precedes PCD in the suspensor, suggesting that the endosperm and suspensor either receive different chemical signals or interpret them differently [119]. A nuclear-localized GA-induced nuclease was found to be active just prior to the appearance of DNA laddering in wheat aleurone cells undergoing PCD. Interestingly, this GA-induced nuclease is not detected in GA-insensitive mutants or when GA synthesis is inhibited [120]. Furthermore, aleurone layers that have not been treated with GAs do not complete PCD.

Foundational biochemical experiments revealed that plant nucleases are localized to a variety of different cellular spaces [5, 116, 120, 121]; for example, in barley aleurone, nuclease activity was found in the ER, Golgi, protein body, and vacuole [122]. A nuclease that is a promising candidate for involvement in PCD is the bifunctional nuclease-1 (BFN1). In *Arabidopsis*, the *BFN1* gene is induced during senescence, abscission, and dPCD [123]. Recent studies revealed that the BFN1 protein, which possesses RNase and DNase activity, is responsible for rapid cell-autonomous corpse clearance and DNA fragmentation during root cap cell death [118, 124]. TUNEL assays showed a delay in nucleic acid degradation in both the nuclei and the cytoplasms of *BFN1* mutants [123]. ORE1, a NAC (ANAC092) transcription factor that positively regulates leaf senescence, has been demonstrated to control the *BFN1* expression [125]. ORE1 is located downstream of the ethylene signaling cascade. Considering this data, it is not surprising that BFN1 is an accepted marker of both plant senescence and PCD. Another nuclease, known as *Zinnia* endonuclease-1 (ZEN1), which shares a number of similarities with BFN1, has been directly implicated to function in PCD. ZEN1 is localized to vacuoles, which collapse before DNA is degraded. ZEN1 was demonstrated to be responsible for nuclear DNA fragmentation during PCD associated with xylem development [126]. Furthermore, silencing *ZEN1* prevented the degradation of nuclear DNA, but did not affect vacuole collapse in a *Zinnia elegans* cell suspension culture. While these findings support the notion that ZEN1 may play a central role in plant DNA fragmentation [126], evidence exists that suggests that multiple nucleases are involved in plant PCD [15].

Given the limited number of nucleases known to be involved in PCD, it has proven to be a challenge to identify these PCD-associated endonucleases. Once identified, other hurdles remain. Based on a study of the role of the nucleases in the process of leaf senescence [127], future studies must explore whether the nuclease is involved in cell death in different tissues in a same plant or in different processes (e.g., fertilization, zygotic embryogenesis, and seed dormancy and germination), is subcellularly localized to the nucleus, or possesses a PCD phenotype—an activity that involves creation of a mutant for the nuclease followed by experiments to observe the impact on genomic DNA during PCD. Additionally, these studies must consider how different nucleases work together to degrade nuclear and organelle DNA.

5. Concluding remarks and future perspectives

Although the biochemical and molecular understanding of plant PCD has increased over the last decade, the mechanisms of action are still very limited and restricted to determine species and some organs and cell compartments. Given its importance, the origin and evolution of genes involved in PCD still need to be resolved. For example, all through seed evolution, PCD has played a fundamental role. In my opinion, eight important goals are key for the best knowledge of plant PCD: (i) the molecular components used in its execution, (ii) the components that have been conserved during evolution, (iii) specific components of PCD, (iv) temporal and spatial expression of Cys-EPs involved in PCD, (v) subcellular Cys-EPs localization and interaction with other proteins, (vi) how the different proteases orchestrate PCD and if there is functional redundancy between the different gene families, (vii) activation of relevant Cys-EPs, and (viii) knowledge of the in vivo protein substrates. An example that justifies what has been said previously can be that VPEs are proposed to control indirectly tonoplast rupture during PCD. However, the detailed mechanism by which VPEs control tonoplast rupture is still diffuse. Together, the intense research carried out in the last decade on PCD in seeds is a strong scientific support to understand the coexistence between death and life.

Acknowledgements

We wish to thank Drs. A. Gunawardena, C. Gietl, S. Maldonado, and M. Nowack for kindly sending their recent manuscripts. I apologize to all those scientists whose manuscripts and/or illustrations have not been directly mentioned.

Conflict of interest

The authors declare no conflict of interest.

Author details

Angel J. Matilla
Department of Functional Biology, Faculty of Pharmacy, University of Santiago de Compostela (USC), Santiago de Compostela, Spain

*Address all correspondence to: angeljesus.matilla@usc.es

References

[1] Rantong G, Gunawardena AHLAN. Programmed cell death: Genes involved in signaling, regulation, and execution in plants and animals. Botany. 2015;**93**:193-210. DOI: 10.1139/cjb-2014-0152

[2] Daneva A, Gao Z, Durme MV, Nowack MK. Functions and regulation of programmed cell death in plant development. Annual Review of Cell and Developmental Biology. 2016;**32**:441-468. DOI: 10.1146/annurev-cellbio-111315-124915

[3] Durand PM, Sym S, Michod RE. Programmed cell death and complexity in microbial systems. Current Biology. 2016;**26**:R587-R593. DOI: 10.1016/j.cub.2016.05.057

[4] Young TE, Gallie DR. Programmed cell death during endosperm development. Plant Molecular Biology. 2000;**44**:283-301

[5] Domínguez F, Cejudo FJ. Programmed cell death (PCD): An essential process of cereal seed development and germination. Frontiers in Plant Science. 2014;**5**:366. DOI: 10.3389/fpls.2014.00366

[6] López-Fernández MP, Maldonado S. Programmed cell death in seeds of angiosperms. Journal of Integrative Plant Biology. 2015;**57**:996-1002. DOI: 10.1111/jipb.12367

[7] Locato V, De Gara L. Programmed cell death in plants: An overview. In: De Gara L, Locato V, editors. Methods in Molecular Biology. New York, NY: Humana Press; 2018;**1743**:1-8. DOI: 10.1007/978-1-4939-7668-3_1

[8] Latrasse D, Benhamed M, Bergounioux C, Raynaud C, Delarue M. Plant programmed cell death from a chromatin point of view. Journal of Experimental Botany. 2016;**67**:5887-5900. DOI: 10.1093/jxb/erw329

[9] Vacca RA, Valenti D, Bobba A, Merafina RS, Passarella S, Marra E. Cytochrome-c is released in a reactive oxygen species-dependent manner and is degraded via caspase-like proteases in tobacco BY-2 cells in route to heat shock-induced cell death. Plant Physiology. 2006;**141**:208-219. DOI: 10.1104/pp.106.078683

[10] Michaeli S, Galili G, Genschik P, Fernie AR, Avin-Wittenberg T. Autophagy in plants—What's new on the menu? Trends in Plant Science. 2016;**21**:134-144. DOI: 10.1016/j.t.plants.2015.10.008.ik

[11] Michaeli S, Galili G. Degradation of organelles or specific organelle components via selective autophagy in plant cells. International Journal of Molecular Sciences. 2014;**15**:7624-7638. DOI: 10.3390/ijms15057624

[12] Koyano T, Kurusu T, Hanamata S, Kuchitsu K. Regulation of vacuole-mediated programmed cell death during innate immunity and reproductive development in plants. In: Sawada H, Inoue N, Iwano M, editors. Sexual Reproduction in Animals and Plants. Tokyo: Springer; 2014;**2014**;431-440. https://doi.org/10.1007/978-4-431-54589-7_36

[13] Senatore A, Trobacher CP, Greenwood JS. Ricinosomes predict programmed cell death leading to anther dehiscence in tomato. Plant Physiology. 2009;**149**:775-790. DOI: 10.1104/pp.108.132720

[14] Höwing T, Dann M, Hoefle C, Hückelhoven R, Gietl C. Involvement of *A. thaliana* endoplasmic reticulum KDEL-tailed cysteine endopeptidase 1 (AtCEP1) in powdery mildew induced and AtCPR5-controlled cell death. PLoS One. 2017;**12**:e0183870. DOI: 10.1371/journal.pone.0183870

[15] Huysmans M, Lema A, Coll N, Nowack MK. Current Opinion Plant Biology. 2017;**35**:37-44. DOI: 10.1016/j.pbi.2016.11.005

[16] Mittler R. ROS are good. Trends in Plant Science. 2017;**22**:11-19. DOI: 10.1016/j.tplants.2016.08.002

[17] Gunawardena AN, McCabe PF. In: Gunawardena AN, McCabe PF, editors. Plant Programmed Cell Death. Cham: Springer; 2015. ISBN: 978-3-319-21033-9

[18] Bruggeman Q, Raynaud C, Benhamed M, Delarue M. To die or not to die?—Lessons from lesion mimic mutants. Frontiers in Plant Science. 2015;**6**:24. DOI: 10.3389/fpls.2015.00024

[19] van Doorn WG, Woltering EJ. Physiology and molecular biology of petal senescence. Journal of Experimental Botany. 2008;**59**:453-480. DOI: 10.1093/jxb/erm356

[20] Thomas H. Senescence, ageing and death of the whole plant. New Phytologist. 2013;**197**:696-711. DOI: 10.1111/nph.12047

[21] Dickman MB, Fluhr R. Centrality of host cell death in plant-microbe interactions. Annual Review of Phytopathology. 2013;**51**:543-570. DOI: 10.1146/annurev-phyto-081211-173027

[22] Matilla MA. Metabolic responses of plants upon different plant-pathogen interactions. In: Ahmad P, Singh VP, editors. Plant Metabolites and Regulation under Environmental Stress. NY: Elsevier-Academic Press; 2018. ISBN: 978-0-128-12689-9

[23] Matilla MA, Krell T. Chapter 3: Plant growth promotion and biocontrol mediated by plant-associated bacteria. In: Egamberdieva D, Ahmad P, editors. Plant Microbiome: Stress Response. NY: Springer; 2018. ISBN: 978-9-811-05513-3

[24] Olvera-Carrillo Y et al. A conserved core of programmed cell death indicator genes discriminates developmentally and environmentally induced programmed cell death in plant. Plant Physiology. 2015;**169**:2684-2699. DOI: 10.1104/pp.15.00769

[25] Gladish DK. Vascular aerenchyma and PCD. In: Gunawardena AHLAN, McCabe PF, editors. Plant Programmed Cell Death. Springer; 2015. pp. 97-121. ISBN: 978-3-319-21033-9

[26] Escámez S, Tuominen H. Programmes of cell death and autolysis in tracheary elements: When a suicidal cell arranges its own corpse removal. Journal of Experimental Botany. 2014;**65**:1313-1321. DOI: 10.1093/jxb/eru057

[27] Iglesias-Fernández R, Matilla A. Flooding stress and O_2-shortage in plants: An overview. In: Ahmad P, editor. Water Stress and Crop Plants: A Sustainable Approach. 1st ed. Vol. 2. John Wiley & Sons, Ltd.; 2016. pp. 711-731. ISBN: 978-1-119-05436-8

[28] Kumpf RP, Nowack MK. The root cap: A short story of life and death. Journal of Experimental Botany. 2015;**66**:5651-5662. DOI: 10.1093/jxb/erv295

[29] Szewinska J, Sminska J, Wielawski W. The roles of cysteine proteases and phytocystatins in development and germination of cereal seeds. Journal of Plant Physiology. 2016;**207**:10-21. DOI: 10.1016/j.jplph.2016.09.008

[30] Schmid M, Simpson DJ, Sarioglu H, Lottspeich F, Gietl C. The ricinosomes of senescing plant tissue bud from the endoplasmic reticulum. Proceeding of the National Academy of Sciences USA. 2001;**98**:5353-5358. DOI: 10.1073/pnas.061038298

[31] Nakano RT, Yamada K, Bednarek P, Nishimura M, Hara-Nishimura I. ER bodies in plants of the Brassicales order:

Biogenesis and association with innate immunity. Frontiers in Plant Science. 2014;**5**:73. DOI: 10.3389/fpls.2014.00073

[32] Hara-Nishimura I, Shimada T, Hatano K, Nishimura M. Transport of storage proteins to protein storage vacuoles is mediated by large precursor-accumulating vesicles. Plant Cell. 1998;**10**:825-836. DOI: 10.1105/tpc.10.5.825

[33] Mollenhauer HH, Totten C. Studies on seeds. V. Microbodies, glyoxysomes, and ricinosomes of castor bean endosperm. Plant Physiology. 1970;**46**:794-799. DOI: 10.1104/pp.46.6.794

[34] Vigil EL. Cytochemical and developmental changes in microbodies (glyoxisomes) and related organelles of castor bean endosperm. The Journal of Cell Biology. 1970;**46**:435-454. DOI: 10.1083/jcb.46.3.435

[35] de Bono AG, Greenwood JS. Characterization of programmed cell death in the endosperm cells of tomato seed: Two distinct death programs. Canadian Journal of Botany. 2006;**84**:791-804. DOI: 10.1139/b06-034

[36] Greenwood JS, Helm M, Gietl C. Ricinosomes and endosperm transfer cell structure in programmed cell death of the nucellus during Ricinus seed development. Proceeding of National of Academy of Sciences, USA. 2005;**102**:2238-2243. DOI: 10.1073/pnas.0409429102

[37] Schmid M, Simpson D, Kalousek F, Gietl C. A cysteine endopeptidase with a C-terminal KDEL motif isolated from castor bean endosperm is a marker enzyme for the ricinosome, a putative lytic compartment. Planta. 1998;**206**:466-475. DOI: 10.1007/s004250050423

[38] Lemos G, Hernández J, Amancio AE, Sales KV. Programmed cell death-related proteases in plants. In: Senturk M, editor. Biochemistry, Genetics and Molecular Biology "Enzyme Inhibitors and Activators". 2017. ISBN: 978-953-51-3058-1

[39] Okomoto T, Minamikawa T, Edward G, Vakharia V, Herman E. Posttranslational removal of the carboxyl-terminal KDEL of the cysteine protease SH-EP occurs prior to maturation of the enzyme. Journal of Biological Chemistry. 1999;**274**:11390-11398. DOI: 10.1074/jbc.274.16.11390

[40] Richau KH, Kaschani F, Verdoes M, Pansuriya TC, Niessen S, Stuber K, et al. Subclassification and biochemical analysis of plant papain-like cysteine proteases displays subfamily-specific characteristics. Plant Physiology. 2012;**158**:1583-1599. DOI: 10.1104/pp.112.194001

[41] Rocha J, Pohl S, Roriz-Fonteles CS. Cloning and gene expression analysis of two cDNA of cysteine proteinase genes involved in programmed cell death in the inner integument from developing seeds of *Jatropha curcas* L. Gene Expression Patterns. 2018;**27**:122-127. DOI: 10.1016/j.gep.2017.12.002

[42] Gietl C, Schmid M. Ricinosomes: An organelle for developmentally regulated programmed cell death in senescing plant tissues. Naturwissenschaften. 2001;**2**:49-58

[43] van Doorn WG. Classes of programmed cell death in plants, compared to those in animals. Journal of Experimental Botany. 2011;**62**: 4749-4761. DOI: 10.1093/jxb/err196

[44] Hierl G, Höwing T, Isono E, Lottspeich F, Gietl C. *Ex vivo* processing for maturation of Arabidopsis KDEL-tailed cysteine endopeptidase 2 (AtCEP2) pro-enzyme and its storage in endoplasmic reticulum derived organelles. Plant Molecular Biology.

2014;**84**:605-620. DOI: 10.1007/s11103-013-0157-6

[45] Misas-Villamil JC, Van der Hoorn RAL, Doehlemann G. Papain-like cysteine proteases as hubs in plant immunity. New Phytologist. 2016;**212**:902-907. DOI: 10.1111/nph.14117

[46] Liu H, Hu M, Wang Q, Cheng L, Zhang Z. Role of papain-like cysteine proteases in plant development. Frontiers in Plant Science. 2018;**9**:1717. DOI: 10.3389/fpls.2018.01717

[47] Lu H, Chandrasekar B, Oeljeklaus J, Misas-Villamil J, Wang Z, Shindo T, et al. Subfamily-specific fluorescent probes for cysteine proteases display dynamic protease activities during seed germination. Plant Physiology. 2015;**168**:1462-1475. DOI: 10.1104/pp.114.254466

[48] Tsujiy A, Tsukamoto K, Iwamoto K, Ito Y, Yuasa K. Enzymatic characterization of germination-specific cysteine protease-1 expressed transiently in cotyledons during the early phase of germination. Journal of Biochemistry. 2013;**153**:73-83. DOI: 10.1093/jb/mvs125

[49] Díaz-Mendoza M, Domínguez-Figueroa JD, Velasco-Arroyo B, Cambra I, González-Melendi P, López-Gonzálvez A, et al. HvPap-1 C1A protease and HvCPI-2 cystatin contribute to barley grain filling and germination. Plant Physiology. 2016;**170**:2511-2524. DOI: 10.1104/pp.15.01944

[50] Christoff AP, Rogerio M. The diversity of rice phytocystatins. Molecular Genetics and Genomics. 2014;**289**:1321-1330. DOI: 10.1007/s00438-014-0892-7

[51] Christoff AP, Passaia G, Salvati C, Alves-Ferreira M, Margis-Pinheiro M, Margis R. Rice bifunctional phytocystatin is a dual modulator of legumain and papain-like proteases. Plant Molecular Biology. 2016;**92**:193-207. DOI: 10.1007/s11103-016-0504-5

[52] Akasofu H, Yamauchi D, Mitsuhashi W, Minamikawa T. Nucleotide sequence of cDNA for sulfhydryl-endopeptidase (SH-EP) from cotyledons of germinating *Vigna mungo* seeds. Nucleic Acids Research. 1989;**17**:6733

[53] Toyooka K, Okamoto T, Minamikawa T. Mass transport of proform of a KDEL-tailed cysteine proteinase (SH-EP) to protein storage vacuoles by endoplasmic reticulum-derived vesicle is involved in protein mobilization in germinating seeds. The Journal of Cell Biology. 2000;**148**:453-464. DOI: 0021-9525/2000/02/453/11

[54] Okamoto T, Shimada T, Hara-Nishimura I, Nishimura M, Minamikawa T. C-terminal KDEL sequence of a KDEL-tailed cysteine proteinase (sulfhydryl-endopeptidase) is involved in formation of KDEL vesicle and in efficient vacuolar transport of sulfhydryl-endopeptidase. Plant Physiology. 2003;**132**:1892-1900. DOI: 10.1104/pp.103.021147

[55] Helm M, Schmid M, Hierl G, Terneus K, Tan L, Lottspeich F, et al. KDEL-tailed cysteine endopeptidases involved in programmed cell death, intercalation of new cells and dismantling of extensin scaffolds. American Journal of Botany. 2008;**95**:1049-1062. DOI: 10.3732/ajb.2007404

[56] López-Fernández MP, Maldonado S. Ricinosomes provide an early indicator of suspensor and endosperm cells destined to die during late seed development in quinoa (*Chenopodium quinoa*). Annual Botany 2013;**112**:1253-1262. DOI: 10.1093/aob/mct184

[57] Hierl G, Vothknecht U, Gietl C. Programmed cell death in Ricinus

and Arabidopsis: The function of KDEL cysteine peptidases in development. Physiologia Plantarum. 2012;**145**:103-113. DOI: 10.1111/j.1399-3054.2012.01580.x

[58] Mikkonen A, Porali I, Cercos M, Ho TH. A major cysteine proteinase, EPB, in germinating barley seeds: Structure of two intronless genes and regulation of expression. Plant Molecular Biology. 1996;**31**:239-254. DOI: 10.1007/BF00021787

[59] Höwing T, Huesmann C, Hoefle C, Nagel MK, Isono E, Hückelhoven R, et al. Endoplasmic reticulum KDEL-tailed cysteine endopeptidase-1 of Arabidopsis (AtCEP1) is involved in pathogen defense. Frontiers in Plant Science. 2014;**5**:58. DOI: 10.3389/fpls.2014.00058

[60] Trobacher CP, Senatore A, Holley C, Greenwood JS. Induction of a ricinosomal-protease and programmed cell death in tomato endosperm by gibberellic acid. Planta. 2013;**237**:665-679. DOI: 10.1007/s00425-012-1780-1

[61] Van Hautegem TT, Waters AJ, Goodrich J, Nowack MK. Only in dying, life: Programmed cell death during plant development. Trends in Plant Science. 2015;**20**:102-113. DOI: 10.1016/j.tplants.2014.10.003

[62] Tuan PA, Kumar R, Rehal PK, Toora PK, Ayele BT. Molecular mechanisms underlying abscisic acid/gibberellin balance in the control of seed dormancy and germination in cereals. Frontiers in Plant Science. 2018;**9**:668. DOI: 10.3389/fpls.2018.00668

[63] Radchuk V, Tran V, Radchuck R, et al. Vacuolar processing enzyme 4 contributes to maternal control of grain size in barley by executing programmed cell death in the pericarp. New Phytologist. 2017;**218**:1127. DOI: 10.1111/nph.14729

[64] Becker C, Senyuk V, Shutov AD, Nong VH, Fischer J, Horstmann C, et al. A storage-globulin-degrading endopeptidase of vetch (*Vicia sativa* L.) seeds, is not involved in early steps of storage protein mobilization. European Journal of Biochemistry. 1997;**248**:304-312. DOI: 10.1111/j.1432-1033.1997.00304.x

[65] Nadeau JA, Zhang XS, Li J, O'Neill SD. Ovule development: Identification of stage-specific and tissue-specific cDNAs. Plant Cell. 1996;**8**:213-239. DOI: 10.1105/tpc.8.2.213

[66] He X, Kermode AR. Proteases associated with programmed cell death of megagametophyte cells after germination of white spruce (*Picea glauca*) seeds. Plant Molecular Biology. 2003;**52**:729-744. DOI: 10.1023/A:1025008117046

[67] Schmid M, Simpson D, Gietl C. Programmed cell death in castor bean endosperm is associated with the accumulation and release of a cysteine endopeptidase from ricinosomes. Proceeding of National of Academy of Science USA. 1999;**96**:14159-14164. DOI: 10.1073/pnas.96.24.14159

[68] Paireder M et al. The death enzyme NbCP14 is a unique papain-like cysteine proteinase with a pronounced S2 subsite selectivity. Archives of Biochemistry and Biophysics. 2016;**603**:110-117. DOI: 10.1016/j.abb.2016.05.017

[69] Zhao P, Zhou XM, Zhang LY, Wang W, Ma LG, Yang LB, et al. A bipartite molecular module controls cell death activation in the basal cell lineage of plant embryos. PLoS Biology. 2013;**11**:e1001655. DOI: 10.1371/journal.pbio.1001655

[70] Hara-Nishimura I, Hatsugai N. The role of vacuole in plant cell death. Cell Death and Differentiation. 2011;**18**:298-1304. DOI: 10.1038/cdd.2011.70

[71] Minina EA, Smertenko AP, Bozhkov PV. Vacuolar cell death in plants: Metacaspase releases the brakes on autophagy. Autophagy. 2014;**10**:928-929. DOI: 10.4161/auto.28236

[72] Jones AM. Programmed cell death in development and defense. Plant Physiology. 2001;**125**:94-97. DOI: 10.1104/pp.125.1.94

[73] Chichkova NV, Kim SH, Titova ES, Kalkum M, Morozov VS, Rubtsov YP, et al. A plant caspase-like protease activated during the hypersensitive response. Plant Cell. 2004;**16**:157-171. DOI: 10.1105/tpc.017889

[74] Zheng H, Staehelin LA. Protein storage vacuoles are transformed into lytic vacuoles in root meristematic cells of germinating seedlings by multiple, cell type-specific mechanisms. Plant Physiology. 2011;**155**:2023-2035. DOI: 10.1104/pp.110.170159

[75] Kroemer G. Twenty one more authors. Classification of cell death. Cell Death and Differentiation. 2009;**16**:3-11. DOI: 10.1038/cdd.2008.150

[76] Salvesen GS, Boatright KM. Caspase-8. In: Barrett AJ, Rawlings ND, Woessner JF, editors. Handbook of Proteolytic Enzymes. 2nd ed. San Diego, California, USA: Academic Press; 2004. pp. 1051-1070. ISBN: 9780123822208

[77] Vorster BJ, Cullis CA, Kunert KJ. Plant vacuolar processing enzymes. Frontiers in Plant Science. 2019;**10**:479. Available from: https://doi.org/10.3389/fpls.2019.00479

[78] Hara-Nishimura I, Hatsugai N, Nakaune S, Kuroyanagi M, Nishimura M. Vacuolar processing enzyme: An executor of plant cell death. Current Opinion in Plant Biology. 2005;**8**:404-408. DOI: 10.1016/j.pbi.2005.05.016

[79] Kumamaru T, Uemura Y, Inoue Y, Takemoto Y, Siddiqui SU, Ogawa M, et al. Vacuolar processing enzyme plays an essential role in the crystalline structure of glutelin in rice seed. Plant Cell Physiology. 2010;**51**:38-46. DOI: 10.1093/pcp/pcp165

[80] Shimada T, Yamada K, Kataoka M, Nakaune S, Koumoto Y, Kuroyanagi M, et al. Vacuolar processing enzymes are essential for proper processing of seed storage proteins in *Arabidopsis thaliana*. The Journal of Biological Chemistry. 2003;**278**:32292-32299. DOI: 10.1074/jbc.M305740200

[81] Hara-Nishimura I. Plant Legumain, asparaginyl endopeptidase, vacuolar processing enzyme. In: Barrett AJ, Rawlings ND, Woessner FJ, editors. Handbook of Proteolytic Enzymes. 3rd ed. London, UK: Academic Press; 2013. pp. 2314-2320. ISBN: 9780123822208

[82] Hatsugai N, Kuroyanagi M, Nishimura M, Hara-Nishimura I. A cellular suicide strategy of plants: Vacuole-mediated cell death. Apoptosis. 2006;**11**:905-911. DOI: 10.1007/s10495-006-6601-1

[83] Hatsugai N, Yamada K, Goto-Yamada S, Hara-Nishimura I. Vacuolar processing enzyme in plant programmed cell death. Frontiers in Plant Science. 2015;**6**:234. DOI: 10.3389/fpls.2015.00234

[84] Radchuk V, Tran V, Radchuck R, Díaz-Mendoza M, Weier D, Fuchs J, et al. Vacuolar processing enzyme 4 contributes to maternal control of grain size in barley by executing programmed cell death in the pericarp. New Phytologist. 2018;**218**:1127-1142. DOI: 10.1111/nph.14729

[85] Tran V, Weier D, Radchuk R, Thiel J, Radchuk V. Caspase-like activities accompany programmed cell death events in developing barley grains. PLoS One. 2014;**9**:e109426. DOI: 10.1371/journal.pone.0109426

[86] Li Z, Yue H, Xing D. MAP kinase 6-mediated activation of vacuolar processing enzyme modulates heat shock-induced programmed cell death in Arabidopsis. New Phytologist. 2012;**195**:85-96. DOI: 10.1111/j.1469-8137.2012.04131.x

[87] Hatsugai N, Kuroyanagi M, Yamada K, Meshi T, Tsuda S, Kondo M, et al. Hara-Nishimura I. A plant vacuolar protease, VPE, mediates virus-induced hypersensitive cell death. Science. 2004;**305**:855-858. DOI: 10.1126/science.1099859

[88] Ge Y, Cai YM, Bonneau L, Rotari V, Danon A, McKenzie EA, et al. Inhibition of cathepsin-B by caspase-3 inhibitors blocks programmed cell death in Arabidopsis. Cell Death Diffentiation. 2016;**23**:1493-1501. DOI: 10.1038/cdd.2016.34

[89] Zhang H, Dong S, Wang M, Wang W, Song W, Dou X, et al. The role of vacuolar processing enzyme (VPE) from *Nicotiana benthamiana* in the elicitor-triggered hypersensitive response and stomatal closure. Journal of Experimental Botany. 2010;**61**:3799-3812. DOI: 10.1093/jeb/erq189

[90] Obara K, Kuriyama H, Fukuda H. Direct evidence of active and rapid nuclear degradation triggered by vacuole rupture during programmed cell death in Zinnia. Plant Physiology. 2001;**125**:615-626. DOI: 10.1104/pp.125.2.615

[91] van Doorn WG et al. Morphological classification of plant cell deaths. Cell Death and Differentiation. 2011;**18**:1241-1246. DOI: 10.1038/cdd.2011.36

[92] del Pozo O, Lam E. Caspases and programmed cell death in the hypersensitive response of plants to pathogens. Current Biology. 1998;**8**:1129-1132. DOI: 10.1016/S0960-9822(98)70469-5

[93] Bozhkov PV, Suárez MF, Filonova LH, Daniel G, Zamyatnin AA, Rodríguez-Nieto S, et al. Cysteine protease mcII-Pa executes programmed cell death during plant embryogenesis. Proceeding of the National Academic of Sciences, USA. 2005;**102**:14463-14468. DOI: 10.1073/pnas.o5o6948102

[94] Kuroyanagi M, Yamada K, Hatsugai N, Kondo M, Nishimura M, Hara-Nishimura I. Vacuolar processing enzyme is essential for mycotoxin-induced cell death in *Arabidopsis thaliana*. Journal of Biological Chemistry. 2005;**280**:32914-32920. DOI: 10.1074/jbc.M504476200

[95] Wang W, Cai J, Wang P, Tian S, Qin G. Post-transcriptional regulation of fruit ripening and disease resistance in tomato by the vacuolar protease SIVPE3. Genome Biology. 2017;**18**:47. DOI: 10.1186/s13059-017-1178-2

[96] Kuroyanagi M, Nishimura M, Hara-Nishimura I. Activation of Arabidopsis vacuolar processing enzyme by self-catalytic removal of an auto-inhibitory domain of the C-terminal propeptide. Plant and Cell Physiology. 2002;**43**: 143-151. DOI: 10.1093/pcp/pcf035

[97] Coll NS, Vercammen D, Smidler A, Clover C, Van Breusegem F, Dangl JL, et al. Arabidopsis type I metacaspases control cell death. Science. 2010;**330**:1393-1397. DOI: 10.1126/science.1194980

[98] Julian I, Gandullo J, Santos-Silva LK, Díaz I, Martínez M. Phylogenetically distant barley legumains have a role in both seed and vegetative tissues. Journal of Experimental Botany. 2013;**64**: 2929-2941. DOI: 10.1093/jxb/ert132

[99] Kumar D, Rampuria S, Singh NK, Shukla P, Kirti PB. Characterization of a vacuolar processing enzyme expressed in *Arachis diogoi* in resistance responses against late leaf spot pathogen, *Phaeoisariopsis personata*. Plant

Molecular Biology. 2015;**88**:177-191. DOI: 10.1007/s11103-015-0318-x

[100] Rojo E, Martín R, Carter C, Zouhar J, Pan S, Plotnikova J, et al. VPE exhibits a caspase-like activity that contributes to defense against pathogens. Current Biology. 2004;**14**:1897-1906. DOI: 10.1016/j.cub.2004.09.056

[101] Albertini A, Simeoni F, Galbiati M, Bauer H, Tonelli C, Cominelli E. Involvement of the vacuolar processing enzyme ϒ-VPE in response of *Arabidopsis thaliana* to water stress. Biologia Plantarum. 2014;**58**:531-538. DOI: 10.1007/s10535-014-0417-6

[102] Gruis DF, Schulze J, Jung R. Storage protein accumulation in the absence of the vacuolar processing enzyme family of cysteine proteases. Plant Cell. 2004;**16**:270-290. DOI: 10.1105/tpc.016378

[103] Nakaune S, Yamada K, Kondo M, Kato T, Tabata S, Nishimura M, et al. A vacuolar processing enzyme, δVPE, is involved in seed coat formation at the early stage of development. Plant Cell. 2005;**17**:876-887. DOI: 10.1105/tpc.104.026872

[104] Endo A et al. Tissue-specific transcriptome analysis reveals cell wall metabolism, flavonol biosynthesis, and defense responses are activated in the endosperm of germinating *Arabidopsis thaliana* seeds. Plant Cell Physiology. 2012;**53**:16-27. DOI: 10.1093/pcp/pcr171

[105] Poncet V et al. The Amborella vacuolar processing enzyme family. Frontiers in Plant Science. 2015;**6**:618. DOI: 10.1074/jbc.M504476200

[106] Kuroyanagi M, Yamada K, Hatsugai N, Kondo M, Nishimura M, Hara-Nishimura I. VPE is essential for mycotoxin-induced cell death in *Arabidopsis thaliana*. Journal of Biological Chemistry. 2005;**280**:32914-32920. DOI: 10.1074/jbc.M504476200

[107] Christoff AP, Turchetto-Zolet AC, Margis R. Uncovering legumain genes in rice. Plant Science. 2014;**216**:100-109. DOI: 10.1016/j.plantsci.2013.11.005

[108] Ingram GC. Dying to live: Cell elimination as a developmental strategy in angiosperm seeds. Journal of Experimental Botany. 2017;**68**:785-796. DOI: 10.1093/jxb/erw364

[109] Lombardi L, Casani S, Ceccarelli N, Galleschi L, Picciarelli P, Lorenzi R. Programmed cell death of the nucellus during *Sechium edule* Sw. Seed development is associated with activation of caspase-like proteases. Journal of Expimental Botany. 2007;**58**:2949-2958. DOI: 10.1093/jxb/erm137

[110] Lima NB, Trindade FG, da Cunha M, Oliveira AE, Topping J, Lindsey K, et al. Programmed cell death during development of cowpea (*Vigna unguiculata* (L.) Walp.) seed coat. Plant, Cell and Environment. 2015;**38**:718-728. DOI: 10.1111/pce.12432

[111] Wan L, Xia Q, Qiu X, Selvaraj G. Early stages of seed development in *Brassica napus*: A seed coat-specific cysteine proteinase associated with programmed cell death of the inner integument. Plant Journal. 2002;**30**:1-10. DOI: 10.1046/j.1365-313X.2002.01262.x

[112] Radchuk V, Weier D, Radchuk R, Weschke W, Weber H. Development of maternal seed tissue in barley is mediated by regulated cell expansion and cell disintegration and coordinated with endosperm growth. Journal of Experimental Botany. 2010;**62**:1217-1227. DOI: 10.1093/jxb/erq348

[113] Zaina G, Morassutti C, de Amicis F, Fogher C, Marchetti S. Endonuclease genes up-regulated in tissues undergoing programmed cell death are expressed during male gametogenesis

in barley. Gene. 2003;**315**:43-50. DOI: 10.1016/S0378-1119(03)00820-5

[114] Moyano L, Correa MD, Favre LC, Rodriguez FS, Maldonado S, López-Fernández MP. Mobilization of reserves in the *Araucaria angustifolia* megagametophyte during germination. Frontiers in Plant Science. 2018;**9**:1275. DOI: 10.3389/fpls.2018.01275

[115] Granot G, Morgenstern Y, Khana A, Rapp YG, Pesok A, Nevo E, et al. Internucleosomal DNA fragmentation in wild emmer wheat is catalyzed by S1-type endonucleases translocated to the nucleus upon induction of cell death. Biochimica et Biophysica Acta. 2015;**1849**:239-246. DOI: 10.1016/j.bbagrm.2014.12.003

[116] Mittler R, Lam E. Characterization of nuclease activities and DNA fragmentation induced upon hypersensitive response cell death and mechanical stress. Plant Molecular Biology. 1997;**34**:209-221. DOI: 10.1023/A:1005868402827

[117] Fath A, Bethke P, Londsale J, Meza-Romero R, Jones R. Programmed cell death in cereal aleurone. In: Lam E, Fukuda H, Greenberg J, editors. Programmed Cell Death in Higher Plants. Dordrecht: Springer; 2000. DOI: 10.1007/978-94-010-0934-8_2

[118] Frendrich M, Hautegem V, Van Durme M, Olvera-Carrillo Y, Huysmans M, Karimi M, et al. Programmed cell death controlled by ANAC033/SOMBRERO determines root cap organ size in Arabidopsis. Current Biology. 2014;**24**:931-940. DOI: 10.1016/j.cub.2014.03.025

[119] Wredle U, Walles B, Hakman I. DNA fragmentation and nuclear degradation during programmed cell death in the suspensor and endosperm of *Vicia faba*. International Journal of Plant Sciences. 2001;**162**:1053-1063. DOI: 10.1086/321922

[120] Domínguez F, Moreno J, Cejudo FJ. A gibberellin-induced nuclease is localized in the nucleus of wheat aleurone cells undergoing programmed cell death. The Journal of Biological Chemistry. 2004;**279**:11530-11536. DOI: 10.1074/jbc.M308082200

[121] Brown PH, Ho THD. Barley aleurone layers secrete a nuclease in response to gibberellic acid. Plant Physiology. 1986;**82**:801-806. DOI: 10.1104/pp.82.3.801

[122] Holstein SEH, Kobert B, Hillmer S, Brown PH, Ho THD, Robinson DG. Subcellular-localization of nuclease in barley aleurone. Physiologia Plantarum. 1991;**83**:255-265. DOI: 101111/j.1399-3054.1991.tb02150.x

[123] Pérez-Amador MA, Abler ML, De Rocher EJ, Thompson DM, van Hoof A, LeBrasseur ND, et al. Identification of BFN1, a bifunctional nuclease induced during leaf and stem senescence in *Arabidopsis*. Plant Physiology. 2000;**122**:169-179. DOI: 10.1104/pp.122.1.169

[124] Farage-Barhom S, Burd S, Sonego L, Mett A, Belausov E, Gidoni D, et al. Localization of the Arabidopsis senescence and cell death-associated BFN1 nuclease: From the ER to fragmented nuclei. Molecular Plant. 2011;**4**:1062-1073. DOI: 10.1093/mp/ssr045

[125] Matallana-Ramirez LP, Rauf M, Farage-Barhorm S, Dortay H, Xue GI, Dröge-Laser W, et al. NAC transcription factor ORE1 and senescence-induced BIFUNCTIONAL NUCLEASE1 (BFN1) constitute a regulatory cascade in Arabidopsis. Molecular Plant. 2013;**6**:1438-1452. DOI: 10.1093/mp/sst012

[126] Ito J, Fukuda H. ZEN1 is a key enzyme in the degradation of nuclear DNA during programmed cell death of tracheary elements. Plant Cell.

2002;**14**:3201-3211. DOI: 10.1105/tpc.006411

[127] Sakamoto W, Takami T. Nucleases in higher plants and their possible involvement in DNA degradation during leaf senescence. Journal of Experimental Botany. 2014;**65**: 3835-3843. DOI: 10.1093/jxb/eru091

Chapter 5

Adaptation of Halophytes to Different Habitats

Milagros Bueno González

Abstract

In recent years, global climate change has been altering environmental (severe drought, soil salinization, irregular precipitation, etc.), around world, decreasing crop yield and upsetting the balance of ecosystems. Nonetheless, a group of plants known as halophytes have the ability to survive and develop in saline soils (wetlands, deserts or temperate zones), may be used in agriculture as a possible alternative to crops (salt-sensitive), as well as for fodder, energy production, medicinal purposes, and desalination of salt-affected areas (phytoremediation). This chapter provides a comprehensive summary of the adaptive strategies used by the annual and perennial halophytes on ecophysiological perspectives, to survive in diverse habitats. The results show a great diverse strategies, such as heteromorphism, seed banks, dormancy, rapid germination, and recovery capacity, from saline shock, favoring the chances of seed survival, although these mechanisms depend on light, moisture, temperature, and the type of salt, in which seeds germinate. In addition, it has been included some molecular, and biochemical aspects, discovered in last years, that might improve our understanding of physiology of these plants. It can conclude that halophytes may be as a possible alternative to ease pressure on cropping systems, restored lands degraded, or confer stress tolerance trough gene transfer.

Keywords: abiotic stress, adaptive strategies, euhalophytes, extremophiles, xerohalophytes

1. Introduction

Soil salinity, a major abiotic stress affecting growth and plant productivity worldwide, constitutes one of the main topics of study in the field of biochemistry, and plant physiology [1, 2]. Salinity exerts various negative effects on germination, as well as on plant growth and development, including water deficit accompanied by reductions in photosynthetic activity, nutritional imbalance, ion toxicity, induction and modulation of plant hormones, and many metabolic changes leading to molecular damage due the production of reactive oxygen species (ROS) [3, 4]. Halophytes are plants that have the availability of survive in saline environment. They account for 1–2% flora of the world, and included both dicots and monocots [5]. Depending of their tolerance and demand for salts, halophytes can be distinguished as obligate and facultative halophytes. Obligate halophytes need salt for growth and development, whereas facultative halophytes can thrive also under strict freshwater conditions and in non-saline habitats, meanwhile that glycophytes (salt-sensitive) include most agricultural crops (**Table 1**) [6]. According to the environmental conditions in which they grow, halophytes are further divided into

Types	Features	Examples
Obligate Halophytes	Growth in saline habitats and stimulating the growth with salt (≥ 200 mM NaCl)	*Arthrocnemum sp.* *Frankenia sp.* *Kochia sp.* *Prosopis sp.*
Facultative Halophytes	Growth with moderate salt concentration but their optimum lies in a salt-free or at least low-salt conditions (≤ 200 mM NaCl)	*Aster tripolium* *Atriplex sp.* *Plantago sp.* *Chenopodium quinoa*
Habitat-indifferent	Growth preferently on salt-free soils, but in salt soils has a better growth compared to glycophytes	*Salsola sp.* *Festuca rubra* *Agrostis stolonifera* *Juncus bufonius*
Glycophytes	Salt-sensitive plants (< 100 mM NaCl)	Most Agricultural crops

Table 1.
Classification of plants according to salinity tolerance.

hydro-halophytes and xerohalophytes. Hydro-halophytes need aquatic conditions, or wet soil (mangroves and salt-marsh species inland or along coastlines). Xerohalophytes survive in dry habitats with saline soils (i.e., arid, semi-arid inlands with unpredictable rainfall) and are considered extremophiles when halophytes germinate and reproduce under high-salinity conditions (≥500 mM NaCl) [7, 8]. On the other hand, it bears emphasizing the widespread use currently made of halophytes as alternatives to glycophytic crops, forages and animal feeds, oilseeds and protein crops, energy crops (biofuels and fuelwood), phytoremediation, medicinal plants and other commercial products [9].

In the course of evolution, halophytes (ephemeral, shrubs, and trees) have developed different mechanisms for regulating growth, development, to ensure their survival in high-salt environments (inland or coastal areas, salt marshes, dunes, and deserts) [7, 10, 11]. Halophytes need anatomical and morphological adaptations such as salt glands, salt bladders (for selective exclusion or accumulation of ions), or development of succulence (dilution of ion concentration) in the plant tissue. Adaptations at the physiological level include: the control of ion uptake, especially Na^+, Cl^- and K^+ by roots and transport to leaves, this maintaining the osmotic balance; the compartmentalization of ions between vacuoles, cytosol, and apoplast through H^+-pump (vacuolar, and plasma membrane); biosynthesis of compatible solutes, and osmoprotectors, (proline, glycine-betaine, sugar, and polyols); the control of stomatal opening and closing; efficiency of the photosynthetic pathway; the synthesis and activation of antioxidant enzymes; the generation of nitric oxide; and the induction and modulation of plant hormones [3, 5].

The adaptations mentioned above, are completed with effective strategies that allow seed germination in different habitats. Most halophytic species belong to the families Amaranthaceae and Poaceae (**Table 2**). Strategies such as seed banks, heteromorphism (based on seed size, weight, and time of production), dormancy, rapid germination, and capacity of recovery of germination from exposure to saline shock, provide seeds multiple opportunities to ensure the continuity of a population, and a plasticity that also provides ecological diversity [7]. It is also important to highlights the role of the seed coat, and endogenous hormones in controlling dormancy and germination [12]. Specifically, salt tolerance is also influenced by genetics and the developmental stage. Germination and flowering are sensitive stages that may be altered by an osmotic effect and/or ion toxicity. Finally, the response of the plant also depends on the type of salt, found in the soil, where the seeds germinate (e.g., NaCl, $NaHCO_3$, and Na_2SO_4, etc.) [13, 14]. In several species of Amaranthaceae

Family	Species
Amaranthaceae (formerly Chenopodiaceae)	*Allenrolfea occidentalis, Arthrocnemum macrostachyum, Atriplex sp., Suaeda sp., Salsola sp., Salicornia, Haloxylon sp., Kalidium sp., Kochia sp.*
Poaceae	*Aeluropus sp., Desmostachya sp.,, Halopyrum sp.,, Sporobolus sp., Urochondra sp.,, Hordeum jubatum, Spartina alterniflora,*
Leguminosae (Fabaceae)	*Arachis glabrata, Desmodium sp., Lathyrus sp. Stylosanthes sp., Trifolium sp., Prosopis sp.*
Asteraceae (Compositae)	*Iva annua, Lasthenia glabrata, Cotula cornopifolia,*
Plumbaginaceae	*Limonium sp.*
Aizoaceae	*Mesembryanthemum crystallinum*
Tamaricaceae	*Tamarix sp.*
Malvaceae	*Kosteletzkya pentacarpos*
Arecaceae	*Nypa fruticans*
Solanaceae	*Solanum chilense*
Rhizophoraceae	*Rhizophora mangle*

Table 2.
Families which present halophytic species.

(formerly Chenopodiaceae), some species show salt-tolerance, while others use a salt-avoidance strategy at germinative level. Typical salt-tolerance is reflected by high germination percentages and/or germination rates under high salinity (e.g., *Haloxylon ammodendron, Suaeda physophora, Anabasis salsa, Suaeda salsa*), while a high percentage of recovery germination after the alleviation of salinity reflects a salt-avoidance strategy (e.g., *Borszczowia aralocaspica*, *Ceratoides latens*, *Bassia dasyphylla*). Other seeds can present both salt-tolerance and salt-avoidance characteristics, or one or the other depending on the type of salt [15].

More information on halophytic species, is available at eHALOPH (http://www.sussex.ac.uk/affiliates/halophytes/), a database of halophytes plants offering data on plant type, life form, ecotypes, maximum salinity tolerance, the presence or absence of salt glands, photosynthetic pathway, antioxidant, secondary metabolites, compatible solutes, habitat, and economic use. This database can serve in the selection of halophytes with applications in phytoremediation, ecological restoration, rehabilitation of degraded ecosystems, and biosaline agriculture [16].

Lastly, as it is unfeasible here to survey the large number of articles on dormancy and germination in halophytes published to date, this chapter summarizes only recent key studies on the mechanisms that halophytes use to germinate in the different habitats around of the world. A section on molecular aspects of some halophyte seeds has also been included.

2. Seed heteromorphism

The most detailed studies on heteromorphism, seed banks, and dormancy have been studied in halophytic species from Amaranthaceae family located between Mongolia and China. Most halophytes present dimorphic and heteromorphic species, many annuals and few perennials (**Table 3**), as an important strategy of

Species	Type of seeds	Habitat
Atriplex spp.	*Brown, Black *Large, small	Desert, Saline soil Inland and coastal marshes
Chenopodium album	*Brown (larger) *Black (smaller)	Light-saline soil
Suaeda spp.	*Brown (larger) *Black (smaller)	Saline soil Salt marshes Salinized desert
Salicornia spp.	*Larger *Small	Inland, saline soil Coastal marshes
Salsola spp.	*Larger with or without winged perianth (WP) *Medium with or without WP *Small with or without WP	Coastal regions Desert Saline soil Cold desert

Table 3.
Heteromorphic seeds species of halophytes under saline conditions.

adapting to habitat variability [17, 18]. One of the most widely studied halophyte species, for its multiple economic and ecological uses (food, forage, bioenergy, medicine, and restoration of salinized or contaminated land) is *Suaeda salsa* L. (Fam. Amaranthaceae). The production of dimorphic seeds helps make the plant more competitive and tolerant to a changeable environment [19, 20]. Some seeds are brown (soft outer seed coat), while other seeds are black (hard and smooth outer seed coat), being able growth in different habitats (i.e., saline inland and intertidal zones) in the Yellow River Delta area (Shandong, China) [19]. Ecophysiological studies have shown that brown seeds absorb water more quickly, and have a higher germination rate than do black seeds, whereas the latter are more sensitive in the absence of light (and remain dormant in light) under high saline concentrations. Black seeds have a lower germination percentage, while brown seeds can germinate under salinity concentrations of up to 600 mM NaCl, regardless of light [20, 21]. In addition, in this last seed was discovered a control of compartmentalization of ions (an accumulation of ions in pericarps cultured with 400 mM NaCl, and a lower amount in embryos), accompanied by rapid germination. This ionic compartmentalization could explain the tolerance of these plants to higher salt concentrations in the intertidal soil and avoid Cl^- and Na^+ toxicity in embryos, during seed development. Also, bracts and the seed coat contribute to ion compartmentalization and prevent excessive accumulation in embryos. On the other hand, brown seeds prove heavier in the intertidal zone (more germination), having a genetic trait that enables tolerance of highly saline environments. Therefore, the production of heteromorphic seeds and different germination characteristics can help mother plants adapt to a fluctuating environment [12].

A more in-depth study has been conducted in the Asian species *Suaeda aralocaspica* (an annual halophyte from the desert) with dimorphic seeds, and C4 photosynthesis pathway [this plant contains two enzymes for the carbon fixation (PEPCK, phosphoenol-pyruvate-carboxylase), and (Rubisco, Ribulose-5-phosphate carboxylase), which provides plants greater efficiency in the carboxylation during the few hours of stomatal opening, due to the high temperatures in the climate zones where this plant lives] [22]. In this species, gene expression has been studied in seeds collected from saline-alkaline sandy soils in the southern margin of Junggar Basin in North China. A series of physiological and biochemical events could take place earlier in the germination of the brown seed as compared to the black seed. A large proportion of genes changed significantly at 3 h in brown seed, as opposed to 8 h in black, after imbibition; it was observed transcriptional changes greater in brown than black seed. However, the different characteristics shown in germination between dimorphic seeds of *S. aralocaspica* are not transferred to the descendants and soon disappear in later seedlings stages, presenting no significant difference in growth, and physiological response, in the descendants with or without salinity [23, 24].

2.1 Seed size

Physiological studies have shown that larger seed size, compared with small, improves the germination percentage under saline conditions [25], and that stress conditions may provide signals, to mother plants, to produce diverse phenotypic plasticity in different environments [12]. Also, brown seeds reportedly exhibit higher amounts of phytohormones: ABA (abscisic acid), IAA (indole-3-acetic acid), and ZR (zeatin-riboside) that black seeds. ABA prompts the accumulation of numerous storage proteins, which contribute in increasing the embryo size and weight in brown seeds under salt stress [26]. On the other hand, the phosphatidylglycerol (PG), a glycerophospholipid, improves salt tolerance in seeds by boosting the production of unsatured fatty acids (allowing greater membrane fluidity). The overexpressed of the *SsGPAT* gene in *Arabidopsis* produced high contents of unsaturated fatty acid and significant salinity tolerance [27]. In young seedlings, that germinated from brown seeds of *S. salsa* as well as from the *Thellungiella halophila* (extremophile that can tolerate saline concentrations up to 700 mM NaCl), PG contents were higher and therefore were related to salt tolerance [28, 29].

2.2 Inorganic ions and organic osmolytes

Inorganic ions facilitate seed germination by improving imbibition (by decreasing seeds water potential), prior to radicle emergence. Xu et al. found that the content in Na^+, K^+, Cl^-, and Ca^{2+} were higher in brown seed compared to black, as well as a higher activity of relative transporters in brown seeds [e.g., vacuolar Na^+/H^+ antiporter (NHX), potassium transporter (HAK), chloride channel protein (CLC), Ca^{2+}/H^+ antiporter, and tonoplast (CAX)], which could explain the better germination rate. These mechanisms are involved in maintaining ionic homeostasis and improve water uptake for seeds during germination under salt stress [30].

Halophytes produce or accumulate more organic osmolytes (proline, betaine, soluble sugar, and polyols) and protein in their large seeds as compared with small seeds to ensure optimum germination under salt stress [31]. These osmolytes can help in osmotic adjustment, can serve as membrane protectors, may aid in ROS detoxification or may act as signaling compounds that trigger other stress-alleviation mechanisms [32]. An osmolyte frequently found in halophytes is betaine. An expression analysis of betaine aldehyde dehydrogenase gene (*SsBADH*) in *S. salsa* showed that brown seeds have higher expression of this gene, and exhibit better germination compared with black seeds. Also, larger brown seeds register a greater increase in sugar facilitating a rapid germination [30].

2.3 Seed coat

Dormant seeds have a hard layer that preserves dormancy until a part of the layer breaks and becomes permeable to water, beginning the process of imbibition. The permeability of the layer in the dimorphic seeds is related to the thickness of the seed layer, which is usually lower in non-latent (tolerant) seeds compared to latent (sensitive) ones. In seeds of *Suaeda physophora*, seed coat play an important role in inhibiting Na^+ influx into and K^+ efflux from the embryo to protect the seed from ion toxicity [33]. In the black seeds of *S. salsa*, a significant accumulation of waxes was detected in the seed coat, in comparison with the brown seeds coat, suggesting that waxes help protect the embryo from ionic toxicity [34]. The waxy substances in the black seeds inhibit water absorption and at the same time maintain the viability of the seeds for longer than in the brown seeds in high-salinity environments [34]. Suberin is one of the major lipids in the seed coat that plays a primitive role in the germination of the seed. Beisson et al. has reported that seed coatings of *Arabidopsis* suberin mutants *jpat5* (Glycerol-3-phospate acyltranferase 5) sharply increase in the suberin accumulation in response to the tetrazolium salt, compared to the wild-type seed coat, reflecting poor seed germination under salt stress [35].

3. Seed banks

In many plants, seeds banks ensure the persistence of populations and contribute to future genetic variability of the species [36, 37]. The studies by Cao et al. in the halophyte *Suaeda corniculata* subsp. *mongolica*, seek to explain the importance of "dormancy" and "seed bank" for the maintenance and regeneration of populations. This plant, an annual herb with succulent leaves, grows in the cold desert of Inner Mongolia and other parts northern China [38, 39]. Seed-bank dynamics, including dormancy cycling of dimorphic seeds, has been studied in detail. The results indicate that black seeds had an annual dormancy/non-dormancy cycle, while brown seeds, remained non-dormant (**Table 4**). Black seeds also exhibited an annual cycle in the sensitivity of germination to salinity. Seedlings that germinated from black seeds emerged in July and August (mid-summer) and those from brown seeds in May (spring). Seedlings were recruited from 2.6% of the black seeds and from 2.8% of the brown seeds in the soil, and only 0.5 and 0.4% of the total number of black and brown seeds in the soil, respectively, gave rise to seedlings that survived to produce seeds. Salinity (NaCl) and water stress induced by polyethylene glycol (PEG) induced dormancy in black seeds and decreased viability of brown seeds. Brown seeds formed only a transient soil-seed bank while black seeds gave rise to a persistent seed bank. Brown seeds germinated under a wider range of

Station	**Black seeds**	**Brown seeds**
Winter: low temperature	Non Dormancy (ND)	ND
Spring: low temperature low precipitation moderate salinity	ND	ND and Germination
Early summer: high temperature high salinity	Conditional dormancy (CD)	Die (85 %)
Mid-summer Late-summer: high temperature high precipitation low salinity	Germination Dormancy	
Autumn: low temperature	CD	
Dormancy cycling	CD →ND →CD→D→CD **Persistent soil seed bank**	**Transient soil seed bank**

Table 4.
Changes of dormancy status of black and brown seeds of Suaeda corniculata, soil seed-bank dynamics, and seedling regeneration of the population.

temperatures and salinities than did the black seeds. Early germination gives plants a selective advantage, since they are less prone to predator and pathogen attack than are late-geminating seeds; they have an early advantage to take up resources, grow much larger, and thus tend to produce more seeds. Germination of black seeds is regulated by an annual dormancy cycle, and these seeds have stricter germination requirements than do brown seeds. That is, black seeds do not germinate in naturally saline habits until soil salinity is decreased by precipitation, and emerge in the summer rainy season. Therefore, the presence of a dormancy cycle in black seeds, but not in brown, and differences in germination requirements of the two dimorphic seeds, cause them to differ in their germination dynamics, ensuring seed availability over a wide range of time [39].

4. Seed dormancy

Dormancy constitutes another defense strategy to govern germination pattern and timing in different habitats, enabling adaptation to environmental changes [40]. The production of dormant seeds (small) remains in soil until the salt concentration becomes conducive to germination. In this scenario, large seeds germinate faster and are more tolerant to salt [41]. In addition, the endogenous content of hormones, ABA (abscisic acid) and gibberellins (gibberellic acid) are keys in this process. These phytohormones regulate dormancy, due their antagonistic functions, GA stimulating and ABA inhibiting this state [42]. However, the endogenous nature of these hormones is not plant species specific [12]. Low germination may be related to higher ABA sensitivity rather than the difference in ABA content among dimorphic seeds in *S. salsa* [43]. Other hormones, such as ethylene, may promote seed germination through the antagonism of ABA signaling, and brassinosteroids may also interact with other hormones to influence seed dormancy by reducing ABA sensitivity and thereby increasing germination [30].

Although it is easier to study annual species with a shorter life cycle, woody species have also been investigated. The first report of a seed-dormancy cycle in a cold-desert shrub is in the halophyte *Kalidium gracile*, a small shrub about 20–50 cm high that grows on alkaline plains and saline lakeshores in northwestern China and Mongolia [44]. Cao et al. have hypothesized that *K. gracile* has a seed bank and dormancy cycling that help restrict germination to a time favorable for seedling survival. Seeds of *K. gracile* were found to have a soil seed bank of 7030 seeds m^{-2} at the beginning of the growing season. About 72% of the seeds were depleted from the soil seed bank during one growing season, and only 1.4% of the seeds gave rise to seedlings that germinated early enough to reach a growth stage at which they could survive the winter. About 28% of the seeds became part of a persistent soil seed bank. Buried seeds exhibited an annual non-dormancy/conditional dormancy (ND/CD) cycle, and germination varied in sensitivity to salinity during the cycle. Dormancy cycling is coordinated with seasonal environmental conditions in such a way that the seeds germinate in summer, when there is sufficient precipitation for seedling establishment. The strategy of *K. gracile* to ensure it survive in these adverse climate zones includes: a polycarpic perennial life cycle, a persistent seed bank, and dormancy cycling. The annual ND/CD cycle in seeds of *K. gracile* contributes to seedling establishment in the unpredictable desert environment and to the maintenance of a persistent soil seed bank [45].

Salsola ferganica (Amaranthaceae), distributed in cold-desert habitats, grows in a heavy saline-alkaline wasteland at the edge of Junggar Basin, Xinjiang (China). This halophyte represents another case of study of heteromorphic seeds, which differ in dispersal ability, dormancy, and germination characteristics. Light could significantly promote germination of their heteromorphic seeds, while GA3 enhances germination, suggesting that *S. ferganica* seeds have a photo-sensitive dormancy type with morphological and non-deep physiological features, light being the dominant factor. In contrast to other desert plant species, *S. ferganica* has comparatively short germinability (only 1–2 years), especially the small seeds, and this is affected by storage time, temperature, salinity, and even the environmental conditions of the maternal plant. It has been reported that both seed polymorphism and seed bank can ensure adequate seedling establishment in unpredictable habitats and consequently promote population propagation. *S. ferganica* has three types of seeds: large sized seeds (LS), middle seeds (MS), and small seed (SS), according to the size of the WP (winged perianth), which has different properties in dispersal ability and germinability. In suitable habitats, the mother plant can produce a large proportion (LP) or middle proportion (MP) of large seeds (LS), or it can produce medium seeds (MS), with a small proportion (SP) of small seeds (SS). They are photo-sensitive species, therefore LP seeds could enter in the potential seed bank, under poor light and unfavorable environments conditions; otherwise, LP seeds would immediately germinate to ensure a large amount of seedling establishment and final population reproduction, under light, and favorable environmental conditions. Therefore, heteromorphic seeds allow *S. ferganica* to gain multiple competitive advantages in unpredictable environments. The seed bank may control the best time for seed germination and seedling establishment, and reduce the risks of spatial and temporal changes of habitats for seed germination, seedling establishment, and population reproduction [46].

5. Germination under saline conditions

Germination is a critical stage in a plant's life cycle, and soil salinity may prevent or inhibit plant development due to osmotic stress and ionic toxicity [7]. As commented above strategies for acclimatization and survival of seeds to various environmental conditions include seed banks, dormancy, and heteromorphism. Nevertheless, germination is the key stage for plant establishment in different habitats [7, 15]. In this section, a number of papers are summarized with regard to the germination of halophytes in different habitats.

The facultative halophyte *Atriplex tatarica* (Amaranthaceae) is a species distributed throughout Central Asia, Asia Minor, south-western Siberia, North Africa and south-eastern Europe (deserts, salt steppes, and disturbed habitats). According to Kochánková and Mandák, seeds germinated as soon as conditions were favorable, thus ensuring short-term survival, but populations risked local extinction when conditions became adverse (i.e., a high-risk strategy). By contrast, germination of the dormant type of seeds was under stronger genetic control and proved to be significantly correlated with basic parameters of population genetics. These seeds ensure long-term reproduction and survival in the field by protracted germination, although in low quantities (i.e., *A. tatarica* also adopts a low-risk strategy), allowing a lasting seed bank over time [47].

Studies on salinity in halophytes are usually carried out with one type of salt; Sosa et al. have investigated *Prosopis strombulifera* (Lam.) Benth (a shrub halophytic), to observe the effect of different types of salt on germination applied alone or mixtures. Seed were collected in southwestern area of the Province of San Luis, Argentina, and the salts used were KCl, NaCl, Na_2SO_4, and K_2SO_4 as well as bisaline iso-osmotic solutions of NaCl + Na_2SO_4 and KCl + K_2SO_4. The germination percentage decreased as salinity increased. In monosaline solutions, SO_4^{-2} with osmotic potentials of −1.2 MPa and lower was more inhibitory than Cl^- at iso-osmotic concentrations. This SO_4^{-2} toxicity was alleviated in salt mixtures and was more noticeable in higher concentrations. Meanwhile, K^+ proved more inhibition than Na^+, independently accompanying anion. Therefore, this study demonstrates that the germination of *P. strombulifera* is strongly influenced by the nature of the ions in the salt solutions and their interactions [48].

In the case of dimorphic seeds, the obligate halophyte *Salicornia europaea* L. has been studied by Orlovsky et al., who used seeds from Central of Asia and provided evidence of high salt tolerance limits. Germination of large seeds was three-to four-fold higher than of small seeds under control and 0.5–2% of all salts tested (NaCl, $Na_2SO_4^{-2}$, 2NaCl + KCl + $CaCl_2$, and $2Na_2SO_4 + K_2SO_4 + MgSO_4$). The germination and plant growth of *S. europaea* in mixed sulfate-chloride salts was distinctly higher than in pure chloride salts, suggesting a mitigating effect of Ca^{2+} and Mg^{2+} ions on the cell-ion balance, reducing the negative impact of Na^+ and Cl^- ions. Small seeds exhibited deep innate dormancy. However, the application of 0.5–2% of chloride and sulfate salts stimulated germination in these seeds, suggesting that they have a different mechanism of salt tolerance than in comparison with large seeds (different mechanism of salt tolerance). In this species, small seeds develop earlier, are more dormants, and prove less salt-tolerant than large seeds. Possibly, small seeds are evolutionarily older, large seeds in *S. europaea* evolving later, when this species habitat had become more saline as a result of species range expansion or increased salinity of the already occupied environment. In short, dimorphic seeds are more flexible in their response to varying salinity, and more adapted to salt and temperature stresses [49].

Song et al. have studied the strategies of adaptation of *Suaeda physophora* (a euhalophyte), and *Haloxylon ammodendron* (a xerohalophyte) during the seed-germination stage, when they were exposed to a range of iso-osmotic NaCl and PEG solutions. The results showed that the fresh weight of germinating seeds was markedly reduced in −2.24 MPa PEG compared with the same amount of NaCl, while the opposite trend was found in the K^+ concentration during the initial 9 h for both species. *H. ammodendron* and *S. physophora* had higher recovery germination from −3.13 MPa NaCl. These responses may produce a persistent seed bank, in saline desert habitats that will maintain the population over time, this representing an important adaptation to saline conditions in one of the driest regions of the world. Another adaptive strategy is for embryo to accumulate less Na^+, although more Na^+ is compartmentalized in the seed coats, which might protect embryos from ion toxicity to ensure seed viability during seed development. In *S. physophora* and *H. ammodendron*, seeds have no endosperm and contain only fully spiral embryo coated by a pericarp; therefore the morphological structure may help these plants germinate as rapidly as possible and may be another adaptive strategy for two species to take advantage of transient favorable conditions during the germination stage. These authors conclude that morphological structure and adaptations to salinity during seed germination may determine the geographical distribution of these species in different saline regions [50]. Later, Zhang et al. studied the impact of salinity on seed germination, chlorophyll content, chloroplast structure and photosynthesis of the green embryos in desiccated seeds of the two species mentioned above. Seeds collected in Fukang Desert Ecological Station of the Xinjiang Institute of Ecology and Geography (China) reveal that the chlorophyll in the cotyledons of desiccated seed had a photosynthetic function in the early germination stage, even under high-salinity conditions. In addition, the photosynthesis in the embryonic cotyledons of desiccated seeds, during germination, was similar to that in the leaves of young seedlings for both halophytes. The photosynthetic function in cotyledons of mature seeds may be ecologically important for seedling development in the early stage for plants growing in extremely saline or arid environments [51].

Different germination rates have been found in two halophytes, showing different salinity tolerance, collected from various salt marshes of southern Spain. *Atriplex prostrata* Boucher ex DC (Amaranthaceae) has early and strong adaptive mechanisms that enable the plant to tolerate exposure from 0 to 200 mM NaCl and maintain a high germination capacity (**Figure 1**), whereas the salinity tolerance of *Plantago coronopus* L. (Plantaginaceae), was increased progressively with the exposure time to salt (**Figure 1**), at lower germination levels. A delay in germination and slow growth are essential for the survival of more sensitive plants in saline environments, allowing the synthesis of compounds for defense against stress; this encourages slow uptake of water from the saline solution, and thereby delayed the time of radicle emergence [52].

To evaluate the salt tolerance mechanisms of Egyptian *Pancratium maritimum* plant (a perennial endangered halophytic species native to the sand coasts to the Mediterranean Sea), Mohamed et al. studied seed germination, mobilization reserve enzymes, and antioxidant enzymes, under saline conditions (0–400 mM NaCl). The results obtained seem to indicate tolerance can come from high levels of esterase, amylase, catalase, and peroxidase activities, and the unique black, spongy, thick seed coat which may act as barrier to saline ionic. In this way, they concluded that this species is a good candidate in their conservation, being able to be used in restoration of coastal area of the Mediterranean Sea [53].

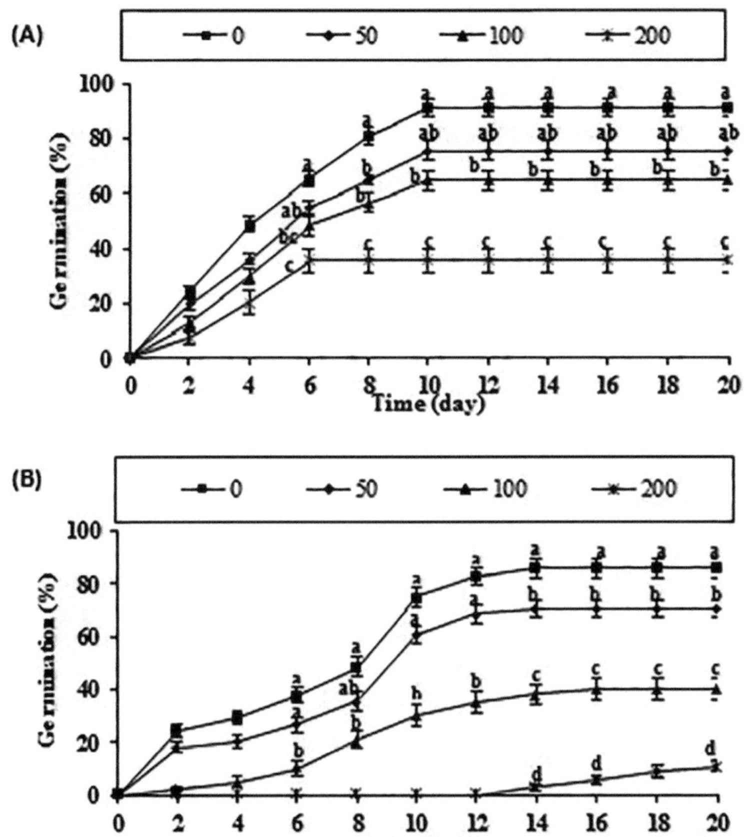

Figure 1.
(A) Germination percentage of Atriplex prostrata seeds treated for 20 days with 0, 50, 100, and 200 mM NaCl. Values presented are mean ± SE (n = 6). Different letters next to the line indicate significant differences between saline treatments (Tukey's test; P < 0.01). (B) Germination percentage of Plantago coronopus seeds treated for 20 days with 0, 50, 100, and 200 mM NaCl. Values presented are mean ± SE (n = 6). Different letters next to the line indicate significant differences between saline treatments (Tukey's test; P < 0.01).

New technologies in study of seeds have helped reveal mechanisms of salinity tolerance in halophytes. Seeds of many coastal plants can survive exposure to seawater and may be dispersed long distances by the ocean. Guja et al. studied natural plant populations in coastal Holocene sand-dune communities near Perth, Western Australia, and collected mature seeds of *Ficinia nodosa* (Cyperaceae) and *Spyridium globulosum* (Rhamnaceae). To determine the response of seeds to external salinity during imbibition, these authors have quantified salt uptake and have resolved the spatially internal distribution of these ones. Flame photometry was used to quantify salt concentration in imbibing seeds and a new application of full-spectrum X-ray mapping enabled the visualization of the spatial distribution and relative abundance of salts. As external salinity increased in *S. globulosum* seeds (salt-sensitive) accumulated sodium and chloride in the seed embryo, while potassium was increasingly displaced and germination was reduced. Conversely, in *F. nodosa* seeds (salt tolerant) avoided ion uptake and germination was not affected by imbibition in high saline concentration (NaCl). These results provide insight into mechanisms of salt tolerance/avoidance during imbibition and early germination, and suggest that oceanic dispersal can be a viable explanation for the distribution of some plant species [54].

5.1 Ions and osmolytes accumulation in halophytes seeds

Prior to germination, the accumulation of osmolytes in seeds grown under saline conditions lowers the water potential of seeds and facilitates imbibition (uptake water from soil) and rapid germination. These osmolytes include free sugars (mannitol, pinitol), proline, betaine, starch, and ions (Na^+, Cl) [14, 55]. Chlorophyll accumulation in dry and imbibed seeds and oxygen production in the embryos of some halophytes may provide energy for the germination stage. In addition, nitrates provided to seeds by maternal plants, may act as signaling molecules to enhance the germination response [20, 56, 57]. Another vital element is calcium. The beneficial effect of Ca^{2+} treatment are attributed mainly to both reducing the rate of Na^+ uptake by roots resulting from blockage of non-selective cation channels (NSCC) by millimolar Ca^{2+} concentrations, as well as to its ability to prevent $NaCl^-$ induced K^+ leak via outward-rectifying channels [58]. Calcium also operates as a second messenger, in both SOS (salt overly sensitive), and ABA signal pathway. Salt-tolerant genotypes appear to have a larger population of Ca^{+2}-sensitive NSCC channels [8, 59].

5.2 Phytohormones in early germination

Phytohormones play a fundamental role in seed germination under saline conditions, with ABA and GA acting as key molecules in the dormancy/germination cycle. The kinetics of ABA and GA during germination in dimorphic seeds of *S. salsa* have been studied in depth. The differential regulation of ABA and GA homeostasis, by salt stress, might help of *S. salsa* plants survive adverse environmental conditions. Li et al. have concluded through morphological analysis that brown and black seeds are at different development stages. Also, ABA accumulation was found in both germinating seed types, with higher induction effect on black than brown seeds, although this gradually decreased after imbibition in water and salt solution. Black seeds showed lower germination percentages than did brown seeds under both water and salt stress, which might be attributed to their higher ABA sensitivity rather than the difference in ABA content between both seeds. Bioactive GA_4 and its biosynthetic precursor reached higher levels in brown than in black seeds, whereas deactivated GAs registered higher contents in black than brown seeds in dry or in germinating (water or salt solutions). The effect of salt stress on GA4 levels in black seeds was stronger than that of brown seeds. Therefore the increased ABA content and sensitivity, as well as the decreased GA_4 content by salinity were more marked in black than brown seeds, contributing to lower germination rates of black seeds exposed to salinity [43].

At the biochemical and molecular level, several genes are strongly induced by NaCl and are involved in the regulation of seed germination through ABA-GA crosstalk during salt stress [60]. A negative regulation of GA and positive biogenesis of ABA is induced by NaCl. Under saline conditions, the ABA level increases several fold due to a stronger expression of genes *ABA-INSENSITIVE* 3 (ABI3) and *ABA-INSENSITIVE 5* (ABI5), which in turn activates the ABA signaling pathway, inhibiting seed germination [61]. Meanwhile, *REPRESSOR OF GA-LIKE 2* (RLG2) transcription is also activated by salinity or by the aBI3/ABI5 pathway, leading to inactivation of the GA signaling pathway, which further inhibits germination by blocking or limiting GA signaling. [8, 60]. In addition, in 22 species of halophytes, ethylene has been found to promote seed germination [62]. Therefore, in halophyte seeds, a small amount of ABA is released, more GA accumulates, and more ethylene synthesis occurs during germination [60, 62].

5.3 Light and temperature in halophytic seeds

Light and temperature, as well as water potential, are key environmental signals that can regulate the time of seed germination, which in turn controls seedling emergence and survival [63]. Agricultural crops tend to germinate in darkness, due to the domestication of species for crop production, whereas halophytic seeds growing in the wild are accustomed to germinating in the light. This is especially critical in the dimorphic seeds, where size may also be related to light. In *S. salsa*, brown seeds may germinate in spring, when soil salinity is high, while black seeds germinate in the latter summer when rainfall can bring them on the soil surface [21]. In *S. corniculata*, rainfall in late summer may bring black seeds to the soil surface, and trigger their germination [39].

Temperature also interacts with salinity and affects seed germination. For example, in *Atriplex rosea*, black seeds are more sensitive to temperature changes [64]. Seed germination in black seeds were found to diminish at lower temperatures regardless of salinity concentration, but brown seeds proved more tolerant to temperature and salinity at cooler conditions (5/15°C). It has been suggested that brown seeds may germinate early in the growing season to preempt the habitat for *A. rosea* [64]. In *Salsola ferganica*, a relatively lower daily temperature range (i.e., 5/15, 10/20, or 15/25°C) could enhance germination of heteromorphic seeds [46]. In the central Asian species *Atriplex*, the optimal temperature regime for black seed germination was 15°C, but 25°C for brown seeds. Moreover, low salinity did not influence the seed germination of black seeds under different temperature regimes (25/35°C), indicating that black seeds can germinate in the rainy summer season. These results imply that the response of dimorphic seeds to combined temperature and salinity could be a major strategy for dimorphic halophytes to survive in changeable saline environments [65].

Salsola imbricata, collected on the northeastern coast of the United Arab Emirates (UAE), is adapted to tolerate higher salinity levels during germination. Elnaggar et al. found high percentages of ready-to-germinate seeds that germinate quickly at lower and moderate temperatures with osmotic potentials of up to −0.8 MPa, indicating that seeds of this species can germinate even in years with less rain than average rainfall, if precipitation occurs early in the growing seasons. Seeds of *S. imbricata* also tolerate lower levels of osmotic potential and show fast germination recovery, especially at lower rather than higher temperatures. These results may indicate the best conditions, especially in dry years, for sowing *S. imbricata* in the restoration of degraded desert sand dunes [66].

6. Recovery of germination

Seed germination recovery ability is a vital adaptive trait for the successful establishment and dispersal of halophytic plants in their native ecosystem. In a Mediterranean climate (warm winters, sunny days year round, and a rainy autumn), seed germination in saline environments usually occurs during spring or a season with high precipitation, when soil-salinity levels are lowered. The capacity of seed dormancy when the soil-water potentials are low, together with the recovery of the germination when the stressful conditions are alleviated, indicate ecological features crucial for the survival of these plants [67]. Pujol et al. studied the recovery of germination in four halophytes of southeastern Spain (the perennials *Halocnemum strobilaceum*, *Arthrocnemum macrostachyum*, *Sarcocornia fruticosa*, and the annual *Salicornia ramosissima*) exposed to is-osmotic stress with different salts (NaCl, $MgCl_2$, $MgSO_4$, and Na_2SO_4) and to an osmotic potential between −2.37 and −3.90 MPa. Pretreatment at low osmotic potentials stimulated the germination velocity, indicating that these species have an apparent ecological advantage since seedlings would become established rapidly before osmotic potentials decreased again. The recovery of germination was similar to that in distilled water (controls), irrespective of the salt used in pretreatment. Therefore, the factor involved in enforcing seed dormancy appears to be the decrease in osmotic potential. The germination ability of perennial species and the annual species does not differ under saline conditions. The osmotic potential of the soils that contain the natural communities promote dormancy in the seeds of the four species and, when the stress conditions are alleviated, germination recovers and the germination rate is stimulated. Although they do produce many seeds, these plants maintain a persistent seed bank [68].

A broader study has been conducted by Shen et al. on six forage species (*Bromus inermis, Elytrigia elongata, Puccinellia tenuiflora, Hordeum brevisubulatum, Kalidium gracile,* and *Suaeda salsa).* The percentage of germination in all the species, except *E. elongata*, significantly declined with rising salinity (0–445 mmol/liter NaCl). The recovery of germination and the length of radicles and shoots, in Hoagland solution, increased with higher salinity. The germination recovery of the six species indicated that germination under NaCl stress was inhibited by osmotic effects. *E. elongata* germination registered the highest NaCl tolerance among six species tested. However, *S. salsa* had low germination even in the control, but salt pretreatment stimulated it to recover germination, which might be associated with the thermoperiod effect. Shoot growth of *P. tenuiflora* had the highest tolerance to NaCl stress, this being the most widely used plant species for saline-soil rehabilitation in northern China [69].

The annual halophyte *Cakile maritima* (Brassicaceae) grows on littoral sand dunes on the semiarid and arid Mediterranean coast of Tunisia. The study of this plant also had double interest, i.e., environmental (for sand-dune fixation) and economic (as seed-oil species). High salinity inhibits and delays germination of *C. maritima* but stimulates seed vigor under recovery. The low osmotic potential generated by high external NaCl concentrations impairs cell division and prevents turgor pressure sufficient to expand and enter in the last phase germination (radicle emergence). During germination, seed-storage proteins (SSPs) are massively broken down, releasing the 20 amino acids that could be transformed into other amino acids. S-adenosylmethionine synthetases (SAM), which are involved in the conversion of cysteine into methionine, were induced during germination, with methionine being present in storage proteins [70]. In control, and especially under salt-stress recovery, serine hydroxymethyltransferase is clearly induced, but to a lesser degree during salt stress. This enzyme converts glycine into serine, which could be further converted into cysteine, this being one of the essential building blocks of glutathione biosynthesis [71]. In short, *C. maritima* avoids germination, under high salinity, by producing quiescent seeds, which can quickly germinate and reach the seedling establishment phase under favorable conditions (when the external water potential increases, from rain). This well-developed phenotypic plasticity, during germination, was better highlighted with the help of proteomic data. The protein profile changes during the germination of *C. maritima*, when storage proteins and fatty acids are massively converted into carbohydrates, slowdown in presence of salt, but are rapidly re-induced upon favorable recovery conditions [67].

7. Certain molecular aspects of halophytes germination

Proteomic, metabolomic, and ionomics studies have been made in many halophytes, in order to understand better the mechanisms of salinity tolerance, focusing the investigation specifically in whole-plants (leaves and roots) [5]. On the other hand, the majority of molecular studies have been made on the model plants *Mesembryanthemum crystallinum* L. (Aizoaceae) and *Eutrema/Thellungiella* spp., (Brassicaceae), a halophyte related to *Arabidopsis,* focusing the study on isolation, gene overexpression, or gene introduction into plants (transgenic plants) [72–74]. However, few studies have been made on seed germination, at the molecular level. Below, a brief summary of recent findings provide a better understanding of salt tolerance of halophytes seeds.

Fukuhara et al. have investigated the latency/germination cycle of seeds of *Mesembryanthemum crystallinum* (L.), called common ice plant [C3-CAM (crassulacean acid metabolism)]. This plant is a system model to study tolerance mechanisms in halophytes. The individual seed capsules spread over time, some seeds germinate within 1 d (early, E) and others do not germinate for more than 4 weeks, until after imbibition (late, L). Several processes such as the uptake of water by the seeds, the start of mitotic activity, and the growth of the radicle seem to be controlled by a mechanism that also establishes different expression patterns of *Vp1* (transcript for transcriptional activator VP1), being VP1 (indicator nature for the threshold that leads from latency to germination). L-seeds are characterized by a lack of expression of the *Cdc2*-related protein (transcript for a cell cycle), and an increase in transcripts of *Vp1*, after water uptake, whereas transcripts related to *Cdc2* increase early and decrease Vp1 at the beginning of E-seeds. Therefore, these authors conclude that the maintenance of the transcription Vp1 may to be the basis of the prolonged latency in the L-seeds. In addition, the expression of several MIP proteins [proteins for aquaporines (AQP) or water channels (WCH)] was characterized during germination and in organs of adult plants. The use of specific probes for the transcription of *Mip (*transcript for AQP and WCH*)* revealed differences during germination that were not due exclusively to water absorption. The *Mip* transcripts increase before the L-seeds begin to germinate. MIP presents greater amounts in active growing tissues. These authors concluded that latency mechanisms for a long time period, before seed germination, provide an advantage for these plants, thus maximizing the survival chances of some seeds in their natural habitat, which is characterized by an unpredictable water supply and consequently environmental stress [75]. Subsequent experiments realized by these authors have clearly shown that the presence and late disappearance of VP1 in L-seeds caused dormancy and after-ripening of L-germinating in these plants [76]. Surówka et al. have realized, an in-depth study, on a transition from C_3 to CAM pathway, suggesting increased temporary of ROS, leading to up-regulation of ROS-responsive genes, catalase (antioxidant enzyme), proline (osmolyte), and polyamines (plant growth regulators). This diurnal rhythm of antioxidant and osmotic compounds could contribute to the maintenance of the water-potential gradient and ROS homeostasis [73].

The extremophile *Eutrema/Thellungiella* spp. is becoming a new model plant because it resembles its relative species (*Arabidopsis thaliana*) in the small genome and the short life cycle. It is highly tolerant to salinity (germination ≤700 mM NaCl) and drought resistant. The high content of several metabolites in these plants indicates metabolic pre-adaptation to salinity to regulate osmotic stress [74]. Little is known about the biochemical and gene-expression changes related to salt, in the germination stage of halophytes. To fill this gap, Kazachkova et al. have shown that germination of *E. salsugineum* is inhibited, by salt, in response to the osmotic component. The seeds of *E. salsugineum* remained viable even when the germination is completely inhibited and the germination resumes once the seeds are transferred to non-saline conditions. In addition, the removal of the seed coat treated with salt allows the embryos to germinate in a medium containing salt. The seeds exposed to salt are also characterized by a reduced proportion of GA/ABA (hormones that control germination and dormancy), and a greater expression of germination repressor genes, *RGL2*, *ABI5*, and *DOG1*. In addition, a salt-mediated increase in the expression of a gene encoding “late embryogenesis abundant” proteins (LEA proteins), and the accumulation of metabolites involved in osmoprotection indicates the induction of processes associated with stress tolerance and the accumulation of easily mobilized carbon stocks. These authors suggest that salt inhibits the germination of seeds by inducing a state with molecular characteristics of dormancy, while the layers of the seed coat provide a physical restriction for the emergence of the radicle. This seed status could facilitate survival in saline soils until a rainfall increases the water potential of soil, indicating favorable conditions for seed germination [77]. On the other hand, halophytes adapted to high levels of salinity have useful genes for improving the salinity tolerance of agricultural crops. One of the growth regulators most related to stress, in plants, are polyamines (putrescine, spermidine and spermine). The *EsPDS1* is a gene that participates in spermidine biosynthesis, specifically encoding the synthesis of Spd-synthase. This gene was cloned and characterized from *E. salsugineum*. The *EsPSS1*-overexpression in tobacco transgenic plant, conferred drought tolerance by alleviating oxidative status of the plant, reflecting the great potential of the study of halophytes [72].

In plants, Halliwell-Asada cycle serve to avoid an excess of ROS, which could damage vital cell components (e.g., AND, membrane lipids, and proteins), and recycles antioxidant systems [78]. Key enzymatic systems include ascorbate peroxidase (APX) an enzyme that eliminates H_2O_2. The function, as well as molecular, and regulatory mechanisms of APX in *Eutrema* (*Thellungiella*) *salsuginea* have unknown. However, Li et al. have studied an APX gene (*TsApx6*), which was cloned from *T. salsuginea*, and its responses in transgenic plants of *Arabidopsis* were analyzed. In the high-salinity treatment, the expression of TsApx6 was significantly induced. In the drought treatment, the overexpression of TsApx6 improved the survival rate and reduced the rate of water loss from the leaves in *Arabidopsis*. Compared with the results for wild-type plants, the high-salinity treatment reduced the concentrations of malondialdehyde (MDA), H_2O_2 (hydrogen peroxide) and proline, but increased the activities of antioxidant enzymes: APX, GPX (glutathione peroxidase), CAT (catalase) and SOD (superoxide-dismutase) in plants overexpressing TsApx6. Meanwhile, the germination rate, the greening of the cotyledon and the length of the root improved in the transgenic plants in comparison

Features	Brown seeds	Blacks seeds	Function in seeds
Size and weight	Heavy and big	Light and small	Different dispersion
Ions (Na^+, K^+, Cl^-)	↑ upregulated	↓ downregulated	Lower water potential Better absorb water
Vacuolar cation/proton exchanger	↑ *CAX3*	↓ *CAX3*	Ion vacuolar compartmentalization
Potassium transport Two-pore potassium channels	↑ upregulated	↓ downregulated	Ion homeostasis
Betaine Aldehyde dehydrogenase *(BADH)*	↑ upregulated	↓ downregulated	Accumulation of betaine (osmolyte) implied in osmotic adjustment
GA 20-oxidase enzyme catalysis steps related to GA biosynthesis	↑ upregulated	↓downregulated	Regulation germination/dormancy
ETR receptor regulates ethylene signaling	↑ upregulated	↓ downregulated	Quick and high germination of brown seeds
Genes related to Brassinosteroids	↑ upregulated	↓ downregulated	Stimulation of germination

Table 5.
Transcriptomic profiling of some genes in maturated dimorphic seeds of Suaeda salsa.

with the wild-type plants, under conditions of salinity and water deficit. These results indicate that, TsApx6 has an important function in plant resistance to certain types of abiotic stress. The bioinformatic analysis indicated that it contains some elements that act in *cis* related to the stress response [79].

In the last years, transcriptomic profiles of some halophytic plants have been carried out by various research groups. The transcriptome dataset permits a better understand of functional genomics in halophytic seeds. Xu et al. realized transcriptomic profiling of genes in matured dimorphic seeds of euhalophyte *Suaeda salsa* (**Table 5**). Upregulated genes involved in seed development, accumulation of osmolytes, and differentially expressed genes of hormones may relate to seed dormancy/germination, and to bigger size and rapid seed germination in brown seeds, compared to black seeds [30].

At molecular level, of halophyte *Karelinia caspica* (Asteraceae), has been studied by Zhang et al. [80]. In addition, the study of differentially expressed genes in this species, under salt stress, and analysis of the effects of salt stress and regulation of ABA could contribute to the identification and characterization of genes and molecular mechanisms underlying saline stress responses, in these plants [80]. *Haloxylon ammodendron*, a perennial xerohalophyte, is an essential species for investigating the effects of drought on desert trees. The results, via RNA-seq and digital gene expression (DGE), were mainly related to ion transporters, signal transduction, ROS-scavenging, photosynthesis, cell wall organization, membrane stabilization and hormones. Moreover, the physiological changes of inorganic ions and organic solute content, peroxidase (POD) activity and osmotic potential were in accordance with dynamic transcript profiles of the more relevant genes. The data provides valuable genetic resources for future improvement of forage and crop species for better adaptation to abiotic stresses [81].

Finally, another widely studied halophyte is *Sporobolus virginicus* (C4 grass) found from tropical to warm temperate regions of world. To identify the key genes involved in the regulation of salt tolerance in this plant, it was produced 3500 independent transgenic *Arabidopsis* lines expressing random cDNA from *S. virginicus* and screened 10 lines which showed enhanced salt tolerance compared with the wild type in a medium containing 150 mM NaCl. Among the selected lines, two contained cDNA coding glycine-rich RNA-binding proteins (*SvGRP1* and *SvGRP2*). GRPs are involved in diverse biological and biochemical processes including salt tolerance in *Arabidopsis* and some other glycophytes. Metabolomic analysis of the *SvGRP1*-transformant suggested that the increase in 3-aminoppropanoic acid, citramalic acid, and isocitric acid content was associated with enhanced salt tolerance [82].

8. Conclusions

This chapter has reviewed the current status of halophytes about the adaptive strategies used to survive in a large heterogeneity of habitats. Phenotypic plasticity, the presence of heteromorphic seeds, persistent and/or transient seed banks, and the cycle of dormancy/germination regulated by ABA and GAs allow ecological diversity, and the most germination appropriate time to achieve success in their survival. The rapid germination and recovery of germination, after drought and saline stress, ensure the propagation of their later generation in areas with varied climatic conditions. These strategies convert halophytes in a viable commercial alternative in countries where the severe climatic conditions do not allow a conventional agriculture, recover degraded lands thanks to phytoremediation action of these plants, and transfer of resistance genes to sensitive agricultural crops, for a better adaptation to abiotic stress. These achievements can help palliate worldwide problems concerning food, ecology, and health, in XXI century.

Author details

Milagros Bueno González
Faculty of Experimental Science, Laboratory of Plant Physiology, Department of Animal Biology, Plant Biology and Ecology, University of Jaén, Jaén, Spain

*Address all correspondence to: mbueno@ujaen.es

References

[1] Golldack D, Li C, Mohan H, Probst N. Tolerance to drought and salt stress in plants: Unraveling the signaling networks. Frontiers in Plant Science. 2014;**5**:151-160. DOI: 10.3389/fpls.2014.00151

[2] Pirasteh-Anosheh H, Ranjbar G, Pakniyat H, Emam Y. Physiological mechanisms of salt stress tolerance in plants: An overview. In: Azooz MM, Ahmad P, editors. Plant-Environment Interactions. Responses and Approaches to Mitigate Stress. 1st ed. Chichester: Wiley; 2016. pp. 141-160

[3] Gupta B, Huang B. Mechanism of salinity tolerance in plants: Physiological, biochemical, and molecular characterization. International Journal of Genomics 2014; 2014:ID701596. DOI: 10.1155/2014/701596

[4] Fahad S, Hussain S, Matloob A, Khan FA, Khaliq A, Saud S, et al. Phytohormones and plant responses to salinity stress: A review. Plant Growth Regulation. 2015;**75**:391-404. DOI: 10.1007/s10725-014-0013-y

[5] Kumari A, Das P, Parida AK, Agarwal PK. Proteomics, metabolomics, and ionomics perspectives of salinity tolerance in halophytes. Frontiers in Plant Science. 2015;**6**:537. DOI: 10.3389/fpls.2015.00537

[6] von Sengbusch P. Halophytes Botanik online. University of Hamburg 2003

[7] Gul B, Ansari R, Flowers TJ, Khan MA. Germination strategies of halophyte seed under salinity. Environmental and Experimental Botany. 2013;**92**:4-18. DOI: 10.1016/envexpbot.2012.11.006

[8] Yuan F, Guo J, Shabala S, Wang B. Reproductive physiology of halophytes: Current standing. Frontiers in Plant Science. 2019;**9**:1954. DOI: 10.3389/fpls.2018.01954

[9] Panta S, Flowers T, Lane PA, Doyle R, Haros G, Shabala SN. Halophyte agriculture: Success stories. Environmental and Experimental Botany. 2014;**107**:71-83. DOI: 10.1016/j.envexpbot.2014.05.006

[10] Flowers TJ, Colmer TD. Salinity tolerance in halophytes. The New Phytologist. 2008;**179**:945-963. DOI: 10.1111/j.1469.8137.2008.02531.x

[11] Flowers TJ, Colmer TD. Plant salt tolerance: Adaptations in halophytes. Annals of Botany. 2015;**115**:327-331. DOI: 10.1093/aob/mcu267

[12] Liu R, Wang L, Tanveer M, Song J. Seed heteromorphism: An important adaptation of halophytes for habitat heterogeneity. Frontiers in Plant Science. 2018;**9**:1515. DOI: 10.3389/fpls.2018.01515

[13] Megdiche W, Amor NB, Debez A, Hessini K, Ksouri R, Zuily-Fodil Y, et al. Salt tolerance of the annual halophyte *Cakile maritima* as affected by the provenance and the development stage. Acta Physiologiae Plantarum. 2007;**29**:375-384. DOI: 10.1007/s11738-007-0047-0

[14] Li X, Zhang X, Song J, Fan H, Feng G, Wang B. Accumulation of ions during seed development under controlled saline conditions of two *Suaeda salsa* populations is related to their adaptation to saline environments. Plant and Soil. 2011;**341**:99-107. DOI: 10.1007/s11104-010-0625-6

[15] Zhang H, Zhang G, Lü X, Zhou D, Han X. Salt tolerance during seed germination and early seedlings stages of 12 halophytes. Plant and

Soil. 2015;**388**:229-241. DOI: 10.1007/s11104-014-2322-3

[16] Santos J, Al-Azzawi M, Aronson J, Flowers TJ. eHALOPH a database of salt-tolerant plants: Helping put halophytes to work. Plant and Cell Physiology. 2016;**57**:1-e10. DOI: 10.1093/pcp/pcv155

[17] Khan MA, Gul B. High salt tolerance in germinating dimorphic seeds of *Arthrocnemum indicum*. International Journal of Plant Sciences. 1998;**159**:826-832. DOI: 10.1086/297603

[18] Yan C, Wei Y, Yang M. Comparative germination of *Tamarix ramosissima* spring and summer seeds. Experimental and Clinical Sciences. 2011;**10**:198-204. PMID: 27857674

[19] Song J, Fan H, Zhao YY, Jia Y, Du XH, Wang BS. Effect of salinity on germination, seedling emergence, seedlings growth and ion accumulation of a euhalophyte *Suaeda salsa* in an intertidal zone and on saline inland. Aquatic Botany. 2008;**88**:331-337. DOI: 10.1016/j.aquabot.2007.11.004

[20] Song J, Wang BS. Using euhalophytes to understand salt tolerance and to develop saline agriculture: *Suaeda salsa* as a promising model. Annals of Botany. 2015;**115**: 541-553. DOI: 10.1093/aob/mcu194

[21] Li W, Liu X, Khan MA, Yamaguchi S. The effect of plant growth regulators, nitric oxide, nitrate, nitrite and light on the germination of dimorphic seeds of *Suaeda salsa* under saline conditions. Journal of Plant Research. 2005;**118**: 207-214. DOI: 10.1007/s10265-005-0212-8

[22] Mao ZM. Flora of Xinjiang. Vol. 2. Urumqi: Xinjiang Science and Technology Press; 1994. pp. 84-106

[23] Cao J, Lv XY, Chen L, Xing JJ, Lan HY. Effects of salinity on the growth, physiology and relevant gene expression of an annual halophyte grown from heteromorphic seeds. AoB Plants. 2015;**7**:plv112. DOI: 10.1093/aobpla/plv112

[24] Cao J, Wang L, Lan H. Validation of reference genes for quantitative RT-PCR normalization in *Suaeda aralocaspica*, an annual halophyte with heteromorphism and C_4 pathway without Kranz anatomy. Peer Journal. 2016;**4**:e1697. DOI: 10.7717/peerj.1697

[25] Wang FX, Yin CH, Song YP, Li Q, Tian CY, Song J. Reproductive allocation and fruit-set pattern in the euhalophyte *Suaeda salsa* in controlled and field conditions. Plant Byosistems. 2018;**152**:749-758. DOI: 10.1080/11263504.2017.1330776

[26] Wang FX, Xu YG, Wang S, Shi WW, Liu RR, Feng G, et al. Salinity affects production and salt tolerance of dimorphic seeds of *Suaeda salsa*. Plant Physiology and Biochemistry. 2015;**95**:41-48. DOI: 10.1016/j.plaphy.2015.07.005

[27] Sui N, Tian SS, Wang WQ, Wang MJ, Fan H. Overexpression of glycerol-3-phospahte acyltranferase from *Suaeda salsa* improves salt tolerance in *Arabidopsis*. Frontiers in Plant Science. 2017;**8**:1337. DOI: 10.3389/fpls.2017.01337

[28] Sui N, Li M, Li K, Song J, Wang BS. Increase in unsaturated fatty acids in membrane lipids of *Suaeda salsa L*. enhances protection of photosystem II under high salinity. Photosynthetica. 2010;**48**:623-629. DOI: 10.1007/s11099-010-0080-x

[29] Sui N, Han GL. Salt-induced photoinhibition of PSII is alleviated in halophyte *Thellungiella halophila* by increases of unsaturated fatty acids in membrane lipids. Acta Physiologiae Plantarum. 2014;**36**:983-992. DOI: 10.1007/s11738-013-1477-5

[30] Xu YG, Zhao YQ, Duan HM, Sui N, Yuan F, Song J. Transcriptomic profiling of genes in matured dimorphic seeds of euhalophyte *Suaeda salsa*. BMC Genomics. 2017;**18**:727. DOI: 10.1186/s12864-017-4104-9

[31] Zhao Y, Yang Y, Song Y, Li Q, Song J. Analysis of storage compounds and inorganic ions in dimorphic seeds of euhalophyte *Suaeda salsa*. Plant Physiology and Biochemistry. 2018;**130**:511-516. DOI: 10.1016/j.plaphy.2018.08.003

[32] Anjum SA, Ashraf U, Tanveer M, Khan I, Hussain S, Shahzad B, et al. Drought induced changes in growth osmolyte accumulation and antioxidant metabolism of three maize hybrids. Frontiers in Plant Science. 2017;**8**:69. DOI: 10.3389/fpls.2017.00069

[33] Song J, Feng G, Li ZK, Chen AD, Chen XM, Zhang FS. Effects of salinity and scarifying seed coat on ion content of embryos and seed germination for *Suaeda physophora* and *Haloxylon ammodendron*. Seed Science and Technology. 2007;**35**:615-623. DOI: 10.15258/sst.2007.35.3.09

[34] Song J, Shi WW, Liu RR, Xu YG, Sui N, Zhou JC, et al. The role of the seed coat in adaptation of dimorphic seeds of the euhalophyte *Suaeda salsa* to salinity. Plant Species Biology. 2017;**32**:107-114. DOI: 10.1111/1442-1984.12132

[35] Beisson F, Li Y, Bonaventure G, Pollard M, Ohlrogge JB. The acyltransferase GPAT5 is required for the synthesis of suberin in seed coat and root of *Arabidopsis*. The Plant Cell. 2007;**19**:351-368. DOI: 10.1105/tpc.106.048033

[36] Bewley JD, Bradford KJ, Hilhorst HWM, Nonogaki H. Environmental regulation of dormancy and germination. In: Bewley JD, Bradford KJ, Hilhorst HWM, Nonogaki H, editors. Seeds. Physiology of Development, Germination and Dormancy. 3rd ed. New York, Heidelberg, Dordrecht, London: Springer; 2013. pp. 299-399. DOI: 10.1007/978-1-4614-4693-4

[37] Anderson TM, Schütz M, Risch AC. Seed germination cues and the importance of the soil seed bank across and environmental gradient in the Serengeti. Oikos. 2012;**121**:306-312. DOI: 10.1111/j.1600-0706.2011.19803-x

[38] Lomonosova M, Brandt R, Freitag H. *Suaeda corniculata* (Chenopodiaceae) and related new taxa from Eurasia. Willdenowia. 2008;**38**:81-109. DOI: 10.3372/wi.38.38105

[39] Cao D, Baskin CC, Baskin JM, Yang F, Huang Z. Comparison of germination and seed bank dynamics of dimorphic seeds of the cold desert halophyte *Suaeda corniculata* subsp. mongolica. Annals of Botany. 2012;**110**:1545-1558. DOI: 10.1093/aob/mcs205

[40] Linkies A, Graeber K, Knight C, Leubner-Metzger G. The evolution of seeds. New Phytologist. 2010;**186**:817-831. DOI: 10.1111./j.1469-8137.2010.03249.x

[41] Wang HL, Wang L, Tian CY, Huang ZY. Germination dimorphism in *Suaeda acuminata*: A new combination of dormancy types for heteromorphic seeds. South African of Journal Botany. 2012;**78**:270-275. DOI: 10.1016/j.sajb.2011.05.012

[42] Kucera B, Cohn MA, Leubner-Metzger G. Plant hormone interactions during seed dormancy release and germination. Seed Science Research. 2005;**15**:281-307. DOI: 10.1079/SSR2005218

[43] Li WQ, Yamaguchi S, Khan MA, An P, Liu XJ, Tran LSP. Roles of gibberelins and abscisic in regulating germination of *Suaeda salsa* dimorphic seeds under salt stress. Frontiers in Plant

Science. 2016;**6**:1235. DOI: 10.3389/fpls.2015.01235

[44] Zhu G, Mosyankin SL, Clemants SE. Chenopodiaceae. In: Wu Z, Raven PH, Hong D, editors. Flora of China 5. Beijing and St Louis: Science Press and Missouri Botanical Garden Press; 2003. pp. 351-414

[45] Cao D, Baskin CC, Baskin JM, Yang F, Huang Z. Dormancy cycling and persistence of seeds in soil a cold desert halophyte shrub. Annals of Botany. 2014;**113**:171-179. DOI: 10.1093/aob/mct256

[46] Ma Y, Wang J, Zhang J, Zhang S, Liu Y, Lan H. Seed heteromorphism and effects of light and abiotic stress on germination of a typical annual halophyte *Salsola ferganica* in cold desert. Frontiers in Plant Science. 2018;**8**:2257. DOI: 10.3389/fpls.2017.02257

[47] Kochánková J, Mandák B. How do population genetic parameters affect germination of the heterocarpic species *Atriplex tatarica* (Amaranthaceae)? Annals of Botany. 2009;**103**:1303-1313. DOI: 10.1093/aob/mcp073

[48] Sosa L, Llanes A, Reinoso H, Reginato M, Luna V. Osmotic and specific ion effects on the germination of *Prosopis strombulifera*. Annals of Botany. 2005;**96**:261-267. DOI: 10.1093/aob/mci173

[49] Orlovsky N, Japakova U, Zhang H, Volis S. Effect of salinity on seed germination, growth and ion content in dimorphic seeds of *Salicornia europaea* L. (Chenopodiaceae). Plant Diversity. 2016;**38**:183189. DOI: 10.1016/j.pld.2016.06.005

[50] Song J, Feng G, Tian C, Zhang F. Strategies for adaptation of *Suaeda physophora*, *Haloxylon ammodendron*, and *Haloxylon persicum* to a saline environment during seed-germination stage. 2005;**96**:399-405. DOI: 10.1093/aob/mci196

[51] Zhang S, Song J, Wang H, Feng G. Effect of salinity on seed germination, ion content and photosynthesis of cotyledons in halophytes or xerophyte growing in Central Asia. Journal of Plant Ecology. 2010;**3**:259-267. DOI: 10.1093/jpe/rtq005

[52] Bueno M, Lendínez ML, Aparicio C, Cordovilla MP. Germination and growth of *Atriplex prostrata* and *Plantago coronopus*: Two strategies to survive in saline habitats. Flora. 2017;**227**:56-63. DOI: 10.1016/j.flora.2016.11.019

[53] Mohamed E, Kasem AMM, Farghali KA. Seed germination of Egyptian *Pancratium maritimum* under salinity with regard to cytology, antioxidant and reserve mobilization enzymes, and seed anatomy. Flora. 2018;**242**:120-127. DOI: 10.1016/j.flora.2018.03.011

[54] Guja L, Wuhrer R, Moran K, Dixon KW, Wardell-Johnson G, Merrit DJ. Full spectrum X-ray mapping reveals differential localization of salt in germinating seeds of differing salt tolerance. Botanical Journal of the Linnean Society. 2013;**173**:129-142

[55] Guo J, Suo S, Wang BS. Sodium chloride improves seed vigour of the euhalophyte *Suaeda salsa*. Seed Science Research. 2015;**25**:335-344. DOI: 10.1017/S0960258515000239

[56] Zhou JC, Fu TT, Sui N, Guo JR, Feng G, Fan JL, et al. The role of salinity in seed maturation of the euhalophyte *Suaeda salsa*. Plant Biosystems. 2014;**150**:83-90. DOI: 10.1080/11263504.2014.976294

[57] Song J, Zhou J, Zhao W, Xu H, Wang F, Xu Y, et al. Effects of salinity and nitrate on production and germination of dimorphic seeds applied both through the mother plant and exogenously

during germination in *Suaeda salsa*. Plant Species Biology. 2016;**31**:19-28. DOI: 10.1111/1442-1984.12071

[58] Shabala S, Cuin TA, Pottosin I. Polyamines prevent NaCl-induced K^+ efflux from pea mesophyll by blocking non-selective cation channels. FEBS Letters. 2007;**581**:1993-1999. DOI: 10.1016/j.febslet.2007.04.032

[59] Chen Z, Pottosin II, Cuin TA, Fuglsang AT, Tester M, Jha D, et al. Root plasma membrane transporters controlling K^+/Na^+ homeostasis in salt-stressed barley. Plant Physiology. 2007;**145**:1714-1725. DOI: 10.1104/pp.107.110262

[60] Yuan K, Rashotte AM, Wysocka-Dillet JW. ABA and GA signaling pathways interact and regulate seed germination and seedlings development under salt stress. Acta Physiologiae Plantarum. 2011;**33**:261-271. DOI: 10.1007/s11738.010-0542-6

[61] Piskurewicz U, Jikumaru Y, Kinoshita N, Nambara E, Kamiya Y, Lopez-Molina L. The gibberellic acid signaling repressor RGL2 inhibits *Arabidopsis* seed germination by stimulating abscisic acid synthesis and ABI5 activity. The Plant Cell. 2008;**20**:2729-2745. DOI: 10.1105/tpc.108.061515

[62] Khan MA, Ansari R, Gul B, Li W. Dormancy and germination responses of halophyte seeds to the application of ethylene. CR Biology. 2009;**332**:806-815. DOI: 10.1016/j.crvi.2009.05.002

[63] El-Keblawy A, Gairola S, Bhatt A. Maternal habitat affects germination requirements of *Anabasis setifera*, a succulent shrub of the Arabian deserts. Acta Botânica Brasílica. 2016;**30**:35-40. DOI: 10.1590/0102-33062015abb0212

[64] Khan MA, Gul B, Weber DJ. Temperature and high salinity effects in germinating dimorphic seeds of *Atriplex rosea*. Western North American Naturalist. 2004;**64**:193-201

[65] Li W, An P, Liu X, Khan MA, Tanaka K. The effect of light, temperature and bracteoles on germination of polymorphic seeds of *Atriplex centralasiatica* Iljin under saline conditions. Seed Science Technology. 2008;**36**:325-338. DOI: 10.15258/sst.2008.36.2.06

[66] Elnaggar A, El-Keblawy A, Mosa KA, Soliman S. Drought tolerance during germination depends on light and temperature of incubation in *Salsola imbricata* a desert shrub of Arabian deserts. Flora. 2018;**249**:156-163. DOI: 10.1016/j.flora.2018.11.001

[67] Debez A, Belghith I, Pich A, Taamalli W, Abdelly C, Braun HP. High salinity impacts germination of the halophyte *Cakile maritima* but primes seeds for a rapid germination upon stress release. Physiologia Plantarum. 2018;**164**:134-144. DOI: 10.1111/ppl.12679

[68] Pujol JA, Calvo JF, Ramírez-Díaz L. Recovery of germination from different osmotic conditions by four halophytes from southeastern Spain. Annals of Botany. 2000;**85**:279-286. DOI: 10.1006/anbo.1999.1028

[69] Shen YY, Li Y, Yan SG. Effects of salinity on germination of six salt-tolerant forage species and their recovery from saline conditions. New Zealand Journal of Agricultural Research. 2003;**46**:263-269. DOI: 10.1080/00288233.2003.9513552

[70] Liu Y, Wu X, Hou W, Li P, Sha W, Tian Y. Structure and function of seed storage proteins in faba bean (*Vicia faba* L.). 3 Biotech. 2017;**7**:74. DOI: 10.1007/s13205-017-0691-z

[71] Noctor G, Mhamdi A, Chaouch S, Han Y, Neukermans J, Marquez-Garcia B, et al. Glutathione in plants: An

integrated overview. Plant, Cell and Environment. 2012;**35**:454-484. DOI: 10.1111/j.1365-3040.2011.02400.x

[72] Zhou C, Sun Y, Ma Z, Wang J. Heterologous expression of *EsSPDS1* in tobacco plants improves drought tolerance with efficient reactive oxygen species scavenging systems. South African Journal of Botany. 2015;**96**:19-28. DOI: 10.1016/j.sajb.2014.10.008

[73] Surówka E, Dziurka M, Kocurek M, Goraj S, Rapacz M, Miszalski Z. Effects of exogenously applied hydrogen peroxide on antioxidant and osmoprotectant profiles and the C_3-CAM shift in the halophyte *Mesembryanthemum crystallinum* L. Journal of Plant Physiology. 2016;**200**:102-110. DOI: 10.1016/j.jplph.2016.05.021

[74] Pin Lee Y, Funk C, Erban A, Kopka J, Köhl KI, Zuther E, et al. Salt stress responses in a geographycally diverse collection of *Eutrema/ Thellungiella* spp. accessions. Functional Plant Biology. 2016;**43**:590-606. DOI: 10.1071/FP15285

[75] Fukuhara T, Kirch HH, Bohnert HJ. Expression of *Vp1* and water channel proteins during seed germination. Plant, Cell and Environment. 1999;**22**:417-424

[76] Fukuhara T, Bohnert HJ. The expression of a *Vp1*-like gene and seed dormancy in *Mesembryanthemum crystallinum*. Genes and Genetic Systems. 2000;**75**:203-209

[77] Kazachkova Y, Khan A, Acuña T, López-Díaz I, Carrera E, Khozin-Goldberg I, et al. Salt induces features of a dormancy-like state in seeds of *Eutrema* (*Thellungiella*) *salsugineum*, a halophytic relative of *Arabidopsis*. Frontiers in Plant Science. 2016;**7**:1071. DOI: 10.3389/fpls.2016.01071

[78] Arashthi H, Pathak RK, Pandey N, Arora S, Mishra AK, Gupta VK, et al. Transcriptome-wide identification of genes involved in Ascorbate–Glutathione cycle (Halliwell–Asada pathway) and related pathway for elucidating its role in antioxidative potential in finger millet (*Eleusine coracana* (L.)). 3 Biotech. 2018;**8**:499

[79] Li Z, Zhang J, Li J, Li H, Zhang G. The functional and regulatory mechanisms of the *Thellungiella salsuginea* Ascorbate peroxidase 6 (*TsAPX6*) in response to salinity and water deficit stresses. PLoS One. 2016;**11**:e0154042. DOI: 10.1371/journal.pone.0154042

[80] Zhang X, Liao M, Chang D, Zhang F. Comparative transcriptome analysis of the Asteraceae halophyte *Karelinia caspica* under salt stress. BMC Research Notes. 2014;7:927. DOI: 10.1186/1756-0500-7-927

[81] Gao HJ, Lü XP, Zhang L, Qiao Y, Zhao Q, Wang YP, et al. Transcriptomic profiling and physiological analysis of *Haloxylon ammodendron* in response to osmotic stress. International Journal of Molecular Sciences. 2018;**19**:84. DOI: 10.3390/ijms19010084

[82] Tada Y, Kawano R, Komatsubara S, Nishimura H, Katsuhara M, Ozaki S, et al. Functional screening of salt tolerance genes from a halophyte *Sporobolus virginicus* and transcriptomic and metabolomic analysis of salt tolerant plants expressing glycine-rich RNA-binding protein. Plant Science. 2019;**278**:54-63. DOI: 10.1016/j.plantsci.2018.10.019

Chapter 6

Assessment of Seed Quality and Germination Response in the Species of the Genus *Polylepis*

Cecilia Vega Krstulovic, Jorge Quezada Portugal, Paola Rocabado Koya, Gabriela Villegas Alvarado and Juan Carlos Bermejo Franco

Abstract

To contribute to the knowledge of sexual reproduction of endangered native species of the genus *Polylepis* (*P. neglecta*, *P. incarum* and *P. pacensis*) in Bolivia, different germination studies were performed to obtain saplings for reforestation as a very important measure recommended for the recovery and conservation of this genus. Initially, seed trees were identified in natural populations (*P. pacensis* and *P. incarum*) and also in green areas of the Municipality of La Paz (*P. neglecta*). The characterization of physical and physiological quality (laboratory and greenhouse tests) of the collected seeds was carried out. The physical quality test, showed a high degree of impurities (47%), so it was required to collect a greater amount of plant material. *P. incarum*, *P. pacensis* and *P. neglecta* showed 10, 8 and 6% humidity respectively. A low germination percentage was obtained: 10% (*P. neglecta*), 8% (*P. pacensis*) and 2% (*P. incarum*), issue reported for several *Polylepis* species. Under greenhouse conditions the highest germination percentage (19%) was displayed in the substrate sand + peat 1:1. On the other hand, *P. neglecta* showed a higher germination percentage (between 10 and 15%), *P. pacensis* and *P. incarum*: 8 and 2%, respectively. Our data suggest feasible alternatives to improve the massive propagation of these species under greenhouse conditions.

Keywords: germination, greenhouse, *Polylepis*, reforestation, seeds, viability

1. Introduction

The arborescent genus *Polylepis* (Rosaceae) is distributed along the Andes mountain range from Venezuela, Colombia, Ecuador, Perú, Bolivia, Argentina (Córdoba), to northern Chile [1].

In Bolivia, it is in an altitudinal range from 3200 to 5200 m, forming true forests in the Eastern Cordillera associated with other species; while in the Western Cordillera it forms monospecific patches [2, 3]. The genus *Polylepis* includes between 15 and 28 recognized species, being Bolivia one of the main centers of wealth with 14 species [3–6].

These forests represent important islands for the conservation of biodiversity in the Andes [7]. They also play a central role in high Andean ecology as a habitat for many species of plants, animals and as an important source of resources for local inhabitants [8]. Nevertheless, because of their high fragmentation, *Polylepis* forests are considered one of the most threatened ecosystems in Bolivia [2, 9].

The degradation of *Polylepis* forests began millennia ago, mainly due to anthropogenic activities such as felling for firewood and charcoal production and human ignited fires to stimulate pastures regrowth in order to enabling agricultural fields and reforestation with exotic species (eucalyptus and pines), factors that have led to the destruction and reduction of ~90% of the area occupied by these forests, restricting them to specialized microhabitats and modifying their floristic and faunal composition [8, 10]. To this must be added the low germination rates (2–50%) reported for different *Polylepis* species, which can be related to the high percentages of empty or non-viable seeds due to the anthropogenic action on their habitat. [1, 11–13].

Consequently, within the framework of the strategy for the conservation of *Polylepis* (Quewiña) forests and biodiversity associated in Bolivia, it was determined that these forested formations are priorities for conservation through the development and implementation of integral projects to restore and recover these ecosystems and their biodiversity, which include the transplantation of juvenile plants and reforestation with native species [14].

In this regard, scientific publications about propagation and reforestation in *Polylepis* are insufficient in Bolivia. So it is essential to get more information about reproductive biology and establishment for this genus.

Nevertheless, there are outstanding experiences in other countries that have generated basic information on different aspects for germination and saplings production for the natural forests recovery. Such as the research carried out in Chile that generated information about *P. tarapacana* seeds collection through the identification of seed trees; seeds germination in a suitable substrate for saplings obtaining; and their transplantation to a definitive place [15]. On the other hand, in Argentina a methodology for the production of *Polylepis australis* seedlings for reforestation has been standardized due to several years of research [12].

In this sense, the study aims to generate information on the sexual reproduction of *Polylepis* species: *P. neglecta*, *P. incarum* and *P. pacensis*, to achieve the understanding of the germinative processes focused on the attributes of physical and physiological quality that can contribute to the generation of standardized treatments to obtain plant material ready to be planted and useful as a tool to implement an integrated management of afforestation and/or reforestation.

2. Methodology

2.1 Selection of species

The species were selected based on the availability of seed trees, that is, *Polylepis* species located in areas near the Department of La Paz-Bolivia were considered, mainly due to the restriction in the geographical area established by the funder to carry out the present investigation. In that sense, the species *P. neglecta*, *P. incarum* and *P. pacensis* were selected.

On the other hand, the state of conservation was considered, of each species that in the case of *P. neglecta* is classified as vulnerable (VU) and *P. pacensis* is within the category endangered (EN) [16]. *P. incarum* qualifies within the category EN [17].

Finally, the growth rate of the species was taken into account, a key aspect for the subsequent production processes of the species of interest. There are reports that *P. neglecta* seedlings show rapid growth (up to 50 cm per year), while *P. incarum* is recognized as a fast-growing tree. Therefore, both species are considered as potential for reforestation [2].

2.1.1 Location and selection of seed trees of Polylepis

In order to verify the zones with forested fragments of *Polylepis*, the sampling points were georeferenced and the selection of potential seed trees was carried out according to specimens that visually presented the highest relative density of fruits in the canopy (presence of branches with fertile structures distributed in the different expositions of the glass). Likewise, basic allometric parameters such as height, trunk diameter DAP_{50} (measured at 50 cm from the base) were recorded, marked and measured and verified by specialists from the National Herbarium of Bolivia (LPB) of the Universidad Mayor de Bolivia. San Andrés (UMSA).

Field trips were made to areas near the Municipality of La Paz, where sectors with *P. pacensis* forests were located, such as the town of Cohoni, at ~75 km from the city of La Paz, between coordinates 16°42′42.9″ S and 67°50′18.4″ W and at an altitude of 3539 m. In it were located different points with presence of fragments of these forests (**Figure 1a**). These populations were settled on slopes and ravines of steep slope; depending on its accessibility, 8 trees were selected and recorded in fruiting state (i.e., presence of immature, mature and remaining fruits), with an average height of 3.7 m and an average DAP_{50} of 17.5 cm.

In the Municipality of Puerto Carabuco, located in the third section of the Camacho Province of the Department of La Paz, at an average altitude of 3860 m, between the coordinates 15°32′17 ″S and 69°25′06″ W, located two main points with presence of fragments of *P. incarum* forests (**Figure 1b**), settled on slopes with steep slopes, where 33 specimens were marked, in which floral and fruit structures were differentiated (mature, immature and old). The average height and DAP50 recorded were 3.5 m and 27.8 cm, respectively.

In the case of *P. neglecta*, a number of previously identified seedbeds were found in the La Paz Botanical Garden (**Figure 1c**), of the University Campus of the Faculty of Pure and Natural Sciences of the UMSA, with the following coordinates 16°32′18″ S and 68°04′7.99″ O, at 3411 m; where 6 trees were marked in flowering and fruiting state. Based on the measurements made, an average height of 3.9 and 26.5 cm of DAP_{50} was calculated.

2.1.2 Seed collection

To collect the seeds of the species under study, trees with mature clusters with a minimum of 25% flowering with respect to the total crown were visually selected, and based on criteria of coloration (fruits of light brown color) and degree of adhesion to the cluster (easily removable). However, in the process, immature (green) seeds were observed that were quite attached to the bunch and old seeds (dark brown) very loose on contact with the hand (**Figure 2**). The collection of seeds was done manually to loosen the fruits. Even so, it was inevitable to obtain samples with different degrees of maturity. The amount of seed collected per tree per species was variable. In the laboratory of the Unit of Plant Biotechnology (U.B.V.-U.M.S.A.) Was homogenized (mixture of primary samples) and reduced to ~100 g in the species *P. neglecta* and *P. incarum*, and only 30 g in the case of *P. pacensis*. From these samples, the distribution of the seeds was carried out according to the quantity required for each of the developed analyzes.

2.2 Analysis of seed quality

Based on the methodology standardized by the International Seed Testing Association [18], for each species under study, the physical and physiological seed quality tests were carried out at the facilities of the U.B.V. of the U.M.S.A.

Figure 1.
(a) Adult individual of P. pacensis, locality Cohoni. Municipality of Palca/detail of leaf, flower and fruit morphology. (b) Adult individual of P. incarum Community of Copacati, municipality of Copacabana/detail of leaf, flower and fruit morphology. (c) Adult individual of P. neglecta Botanical Garden La Paz-University Campus of the U.M.S.A./Detail morphology of leaf, flowers and fruits.

2.2.1 Physical quality

The physical quality analysis included the determination of parameters such as physical purity, moisture content and the weight of 1000 seeds. The physical purity analysis consisted in examining a sample of 100 g of seeds per species (with the exception of *P. pacensis* for not obtaining the sample size) and separating the pure seed from the impurities (seed of other species and inert matter) with the aid

Figure 2.
Seed collection process of Polylepis: (a) seed collection manual; (b) seeds collected in bags; (c) preservative Polylepis seeds; (d) immature seeds; (e) mature seeds; and (f) seeds (old).

of magnifying glasses, to then weigh them and calculate the percentage of purity according to the following formula [19]: purity percentage = (pure seed/total weight of the original sample) × 100.

Subsequently, we proceeded to determine the moisture content of the seed through the method recommended by the ISTA [18]. Eight samples of 100 seeds were weighed, dried in an oven at a temperature of 103 ± 2°C for 17 hours and then weighed again. The humidity percentage was calculated from the following formula: humidity percentage = (original weight − weight after stove drying/original weight) × 100.

For the evaluation of the weight of 1000 seeds, these were separated manually, with 8 repetitions randomly per species and each of them was weighed in an analytical balance.

2.2.2 Physiological quality

An exploratory study was previously carried out using the tetrazolium viability test based on staining patterns that allowed the reliable evaluation of the viability of the seeds of the species under study. In accordance with the procedure established by the ISTA [18], the seeds were pre-conditioned, placed for 24 h on moistened paper towel. This triggered the activation of the enzymatic system, besides facilitating the cutting of the seed and the development of a sharper and uniform coloration. Next, a cross section of the seeds was made to expose the embryo directly to the action of the tetrazolium and facilitate its penetration. Five repetitions of 20 seeds were made per Petri dish, making a total of 100 seeds per species. Subsequently, the seeds were stained by immersing them in the tetrazolium solution at a concentration of 0.5% for a period of 24 h in the dark and at room temperature. To proceed with the interpretation of the test, a stereoscope was used. In this way, the seeds were classified, according to the colored zones, in different staining categories established according to the ISTA standards: Category 1: living tissue (TV1, intense red tissue staining, TV2, bright or intense pink tissue staining, TV3, light pink tissue staining); Category 2: deteriorated tissue (TD1, garnet red tissue coloration. TD2, milky pink tissue coloration); Category 3: dead tissue (TM1, discoloration of discolored/whitish or hyaline tissue TM2, necrotic tissue with brown coloration). Given that empty seeds were found (without an embryo), Category 4 (SSTE, seed without embryonic tissues) was added to describe this condition.

The main attribute considered in the physiological quality of seeds refers to their germination capacity. In this sense, a germination test was carried out under controlled environmental conditions of humidity, temperature and light in the growth room of the U.B.V., which has shelves with artificial light, programmed with a photoperiod of 16 h light/8 dark and an average temperature of 25°C. Six seeds were planted per Petri dish, with a total of 240 seeds for each species. The number of repetitions (Petri dish) per species was 40. As a substrate paper, towel was used, which provided adequate porosity for moisture retention. This trial lasted 30 days, time that was waited for the maximum germination to occur. During this period, the count of (1) germinated seeds (those that developed a seedling with essential structures, such as roots, stems and cotyledons) and (2) ungerminated seeds.

2.3 Seed germination in greenhouse

Considering the information available in previous works on *Polylepis* seeds, two trials were carried out in the greenhouse of the La Paz Botanical Garden-U.M.S.A. In a first (preliminary) trial, four combinations of substrates were tested: S1, sand + peat (1:1) (control substrate); S2, sand + peat + black earth (1:1:1); S3, sand + peat + humus (2:1:1); and S4, sand + black earth (1:1). Prior to planting the seeds, pregerminative soaking in water was carried out for 24 h. Two hundred and forty seeds were sown per species; half was subjected to this procedure. The study was conducted in July, August and September 2015. The data obtained were analyzed statistically through a completely randomized design with trifactorial arrangement; the factors were: Species (A), soaking (B) and substrates (C), with 10 repetitions per treatment. Then a second (final) test was carried out, in which the two best substrates from the preliminary test (S1 [sand + peat 1:1] and S4 [sand + black earth 1:1]) were used to verify the observed response in the previous study and establish treatments for subsequent production processes. In both cases, the pregerminative treatment (seed soaking) was applied. Two hundred and forty seeds per species were planted, of which 120 were soaked in water to standardize a technique for the propagation of keñua seedlings. This study was carried out in the months of September and October 2015. The results were analyzed through a completely randomized design with a bifactorial arrangement; the factors were species (A) and treatments of substrates with and without soaking (B), with 40 repetitions per treatment.

3. Results

3.1 Analysis of seed quality

3.1.1 Physical quality

The results regarding the parameter of physical quality purity of seeds showed that in 100 g of seed of *P. neglecta*, 57% corresponded to pure seed and, in the case of *P. incarum*, 50%. As regards *P. pacensis*, this information was not generated since only a maximum of 30 g of the 100 g required for this analysis was reached. With respect to the moisture content of the seeds, in the case of *P. neglecta* species 6%, 8% for *P. pacensis* and 10% for *P. incarum* were recorded. The average values obtained for the parameter weight of 1000 seeds for each species showed that 1000 seeds of *P. neglecta* weigh 4.78 g, in the case of *P. incarum* 9.65 g and in the species *P. pacensis* 5.31 g.

3.1.2 Physiological quality

Depending on the results obtained in the tetrazolium test, different coloring patterns could be differentiated in the tissues of the seeds. In the species *P. neglecta* 40% of seeds were observed with intense red tissue staining (TV1), 22% with dead tissue (TM2) and 17% of empty seeds (without embryo). In *P. incarum* 48% of seeds with brown coloration (TM2) and 20% with discoloration of colorless/whitish or hyaline tissue (TM1) was verified, condition that indicates tissue death in both cases. In the species *P. pacensis* 58% of seeds with dead tissue (TM2) and 21% of empty seeds were recorded, without the formation of embryonic tissues (SSTE). The rest of the percentages presented in smaller proportion in the three species correspond to other categories that are shown in **Table 1**.

In turn, the different staining categories established to determine the viability of the seeds can be observed in **Figure 3**.

In general, this test allowed us to obtain an approximation to determine the viability of *Polylepis* seeds through the differentiation of colorations of the observed tissues. Giving as final results a high percentage of non-viable seeds in the species *P. incarum* and *P. pacensis* (76 and 85% respectively), while in *P. neglecta* a lower percentage of non-viable seeds was recorded (42%), results that to some extent support the low germination rates observed in both species described below (**Figure 4**).

Species	Category 1 Living tissue			Category 2 Deteriorated tissue		Category 3 Dead tissue		Category 4 Empty seed	Total seeds evaluated
	TV1	TV2	TV3	TD1	TD2	TM1	TM2	SSTE	
P. neglecta	40	16	2	0	0	3	22	17	100
P. incarum	6	10	8	0	0	20	48	8	100
P. pacensis	9	2	4	0	1	5	58	21	100

TV1: intense red tissue coloration. TV2: bright or intense pink tissue coloring. TV3: coloration of light pink tissue). TD1: coloration of garnet red tissue. TD2: milky pink tissue coloration. TM1: coloration of discolored / whitish or hyaline tissue. TM2: necrotic tissue (brown coloration). SSTE: empty seed without embryonic tissues.

Table 1.
Percentage of seeds according to the categories of tetrazolium staining (TZ).

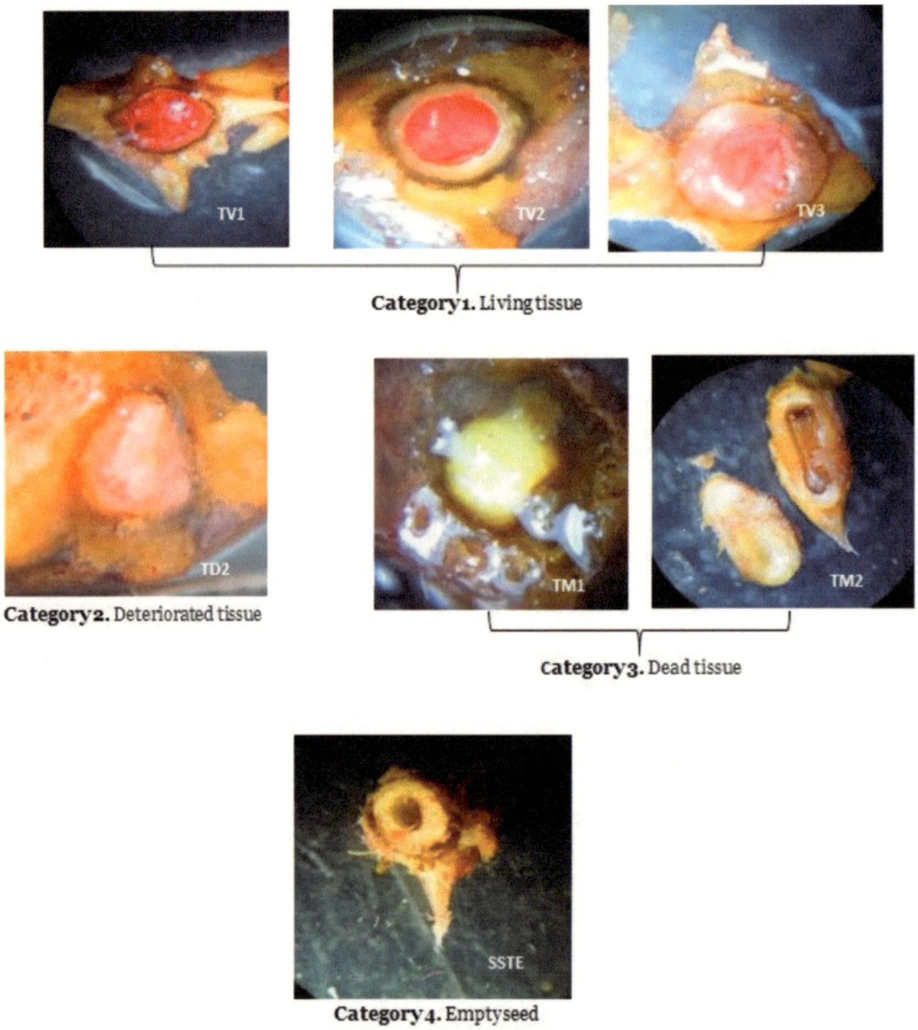

Figure 3.
Colored seeds with tetrazolium, distributed among living, deteriorated, dead tissues and empty seeds.

The results of the germination test showed a germination rate of 42% for *P. neglecta* and 10% for *P. incarum* and *P. pacensis*, after 30 days of evaluation, defined as the observation period in which the maximum occurred accumulated germination (**Figure 5a–c**).

3.1.3 Seed germination in greenhouse

The results of the preliminary test, at 45 days of evaluation (maximum germination time), showed only significant statistical differences with respect to the factors species (A) and substrates (C), not so for the soaking factor (B), nor for any of the interactions between factors. Accordingly, both factors were analyzed separately through Duncan's multiple range test ($\alpha = 0.05$), which he identified as the highest germination percentage to that registered for *P. neglecta* (15%), since it is statistically different from those obtained for *P. incarum* and *P. pacensis* (**Figure 6a**).

Regarding the differences identified for the substrate factor, through the Duncan test ($\alpha = 0.05$) it was verified that the substrate S1 (sand + peat 1:1) had the highest germination rate (19%) in relation to the other substrates (**Figure 6b**). In this sense, when

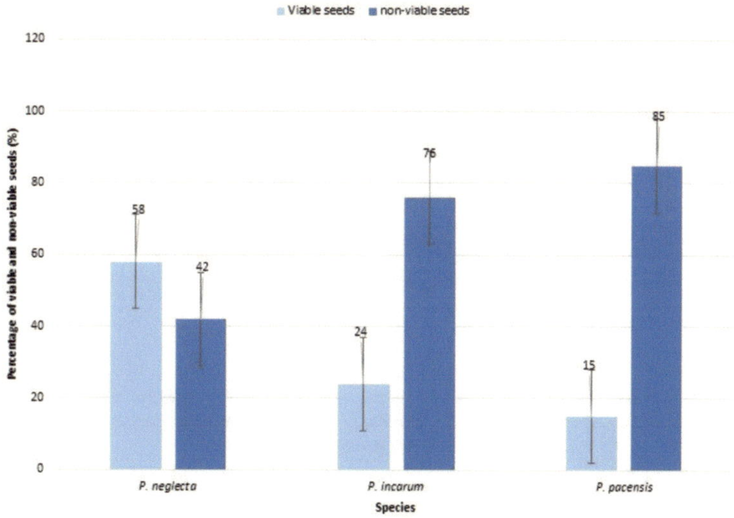

Figure 4.
Percentage of viable and non-viable seeds of Polylepis obtained through the tetrazolium test (TZ).

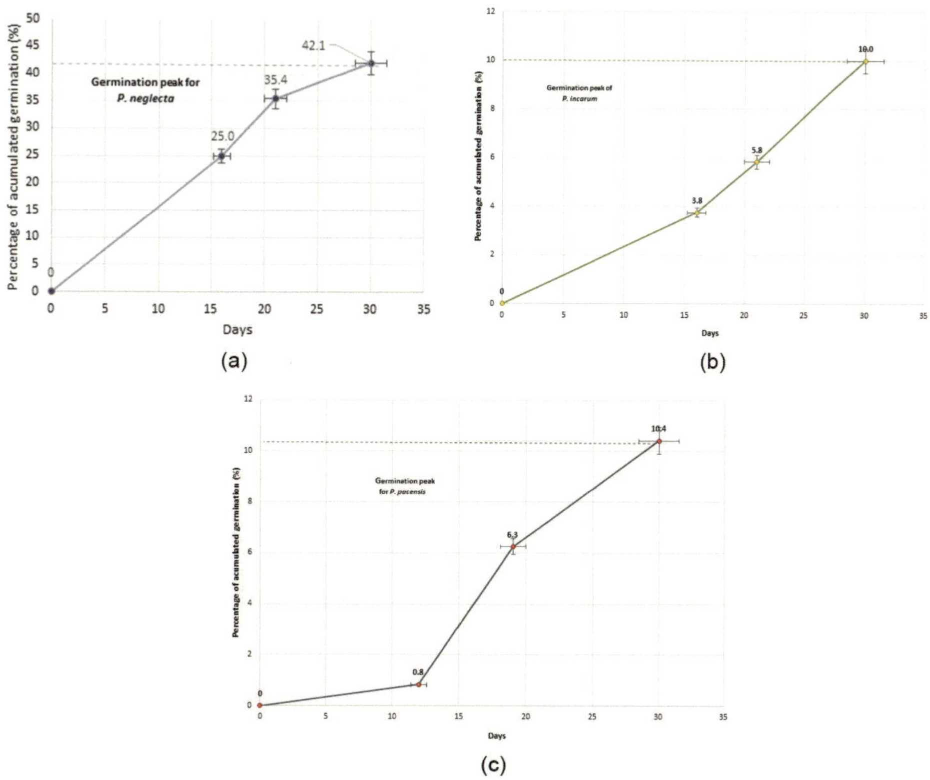

Figure 5.
Germination rate of Polylepis species up to 30 days: (a) P. neglecta, (b) P. incarum and (c) P. pacensis.

dealing with evaluation factors it can be affirmed that the substrate S1 is the one that presents the best germination rate for all the species under study and the species with the best germinative behavior regardless of the substrate or the soaking is *P. neglecta*.

In the final test the results showed significant statistical differences with respect to the species under study (A) and the treatments (B), and not so for the interaction (AB).

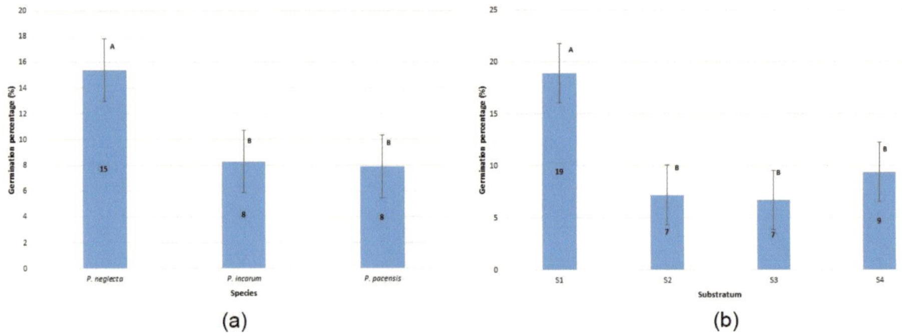

Figure 6.
*Seed germination results of Polylepis species under greenhouse conditions (preliminary test). (a) Percentage of average germination per type of substrate. (b) Percentage of average germination registered by species. *According to Duncan (α = 0.05). Different letters are statistically different.*

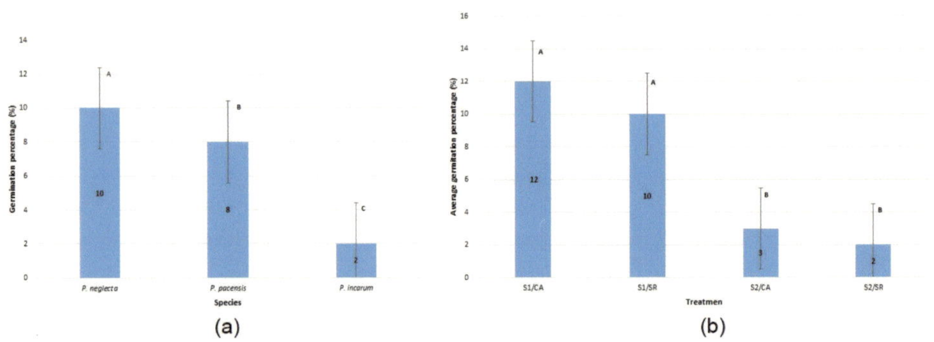

Figure 7.
*Seed germination results of Polylepis species under greenhouse conditions (final test). (a) Percentage of germination registered according to the type of substrate and pre-germinative treatment. (b) Percentage of germination registered by species. *According to Duncan (α = 0.05). Different letters are statistically different.*

In this way, the Duncan test (α = 0.05) again identified *P. neglecta* as the species that reached the highest percentage of germination (10%), followed by *P. pacensis* (8%) and finally *P. incarum* (2%) (**Figure 7a**). Likewise, this test revealed that the treatments with the highest percentage of germination correspond to the substrate S1 (sand + peat 1:1), independently of the effect of the pregerminative soaking treatment (**Figure 7b**). In summary, it can be confirmed that the S1 substrate, independently of the soaking, was the one that presented the best germination rate for the three species. The best germinative behavior, regardless of the substrate or soaking, was the species *P. neglecta*.

4. Discussion

4.1 Analysis of seed quality

With respect to physical quality parameters, in the case of physical purity, the seeds presented a high degree of impurities (47% on average), which makes it difficult to obtain pure seeds for future planting. Its importance lies in the fact that a lot of seeds with a lot of inert material is of lower quality and implies the need to use more of this material in order to make the process more effective [20].

The moisture content is one of the most important factors that affects the maintenance of the quality of the seeds. Thus, dry and healthy seeds can be kept under appropriate storage for many more years, while wet seeds can deteriorate in a few days. The storage time of seeds decreases as the moisture content increases, since it has a dominant effect on the predominance and activity of insects and fungi during storage [21]. The moisture content affects the metabolic activity of the seeds in the germination and the deterioration of these; therefore, knowledge of this attribute allows choosing the most appropriate procedure for the harvest or collection of seeds, their conditioning, the preservation of the physical, physiological and sanitary quality of the seed [20]. In the case of the species studied, there are no recommended humidity parameters for storage. However, according to the ISTA [18] it is considered as a norm that the humidity of the seeds to be commercialized should not exceed 13% [20]. The percentage of humidity of all the species under study was low, which could be considered orthodox seeds, which are those whose moisture content can be reduced to values between 5 and 10% and store them at temperatures below zero without damaging them; its conservation for long periods without losing its germinative power [22]. According to the document published by the Mining Company Doña Inés de Collahuasi SCM [15], seeds of *P. tarapacana* stored for a year in a dry cellar at room temperature did not lose their viability, so the percentage of germination was not seen affected.

The weight of 1000 seeds allows to establish the necessary amount of seeds to achieve a certain number of plants. If the value of this parameter is known, it is easy to determine the amount of seeds to be used per reforestation area [21]. According to what was observed in the present study, it can be inferred that to have 1000 seeds of *P. neglecta* and *P. pacensis*, between 4 and 6 g is required, while in *P. incarum* about 9 g would be required.

In relation to the physiological quality analyzes, through the viability test of tetrazolium in *Polylepis* seeds, high percentages of non-viable seeds and empty seeds were verified, which would be related to the low percentage of germination registered in different species of the *Polylepis* genus. In this regard, it is mentioned that in the collection of seeds corresponding to *P. incana*, many trees came to have almost all their seeds empty [11]. Another very interesting result is the one reported for *P. australis* (Córdoba-Argentina), in which they analyzed the effect of the seed mass, the geographic region, the forest fragments and the individual trees on the germination of the seeds, and reported that seeds that did not germinate were almost always empty [23]. Likewise, these authors refer to Hensen (unpublished data), who found a high percentage of empty and non-viable seeds in several Bolivian *Polylepis* species (*P. hieronymi*, 80%, *P. tomentella*, 84%, *P. besseri*, from 90 to 100%, *P. racemosa*, 100%, *P. tarapacana*, 100%). The presence of high percentages of empty seeds seems to be a characteristic in *Polylepis* species, which represents a problem when making afforestation or reforestation plans. Likewise, the existence of a positive correlation between the mass and seed germination of *P. australis* was determined, that is, the weight of the seed could be a quick way to evaluate the viability of these and, therefore, it would be a practical alternative for the selection of seeds and ensure higher gemination rates [23].

Based on the germination test carried out, available reports recorded low germination rates in the species studied, which was verified through the data obtained for this study. Experiences of germination of *P. incana* in Petri dishes from samples taken at different altitudes showed germination percentages between 8 and 34% in a period of 15–49 days (collection of seeds at 3100 m). Likewise, samples collected at an altitude of 3600 m, with pregerminative treatment consisting of soaking in water for 42 h prior to sowing, showed germination percentages between 12 and 44% [24].

4.2 Seed germination in greenhouse

Different experiences in germination for the production of seedlings in other *Polylepis* species under greenhouse conditions, report results similar to those presented in this chapter. Regarding the type of substrate, they recommend a mixture of sand + peat in a 1:1 ratio, which in the tests carried out in the present study turned out to be the substrate in which the greatest number of seeds of the three species analyzed [2] germinated. Although some researches recommend to subject the seeds to a soaking in water for 24 h as a pregerminative treatment [25, 26], the results in this work did not show a marked effect on the germination of the species under study, so their use resulted indistinct depending on the data obtained.

With relationship the time of germination, the first seedlings emerged between 20 and 25 days in (*P. neglecta*, *P. incarum* and *P. pacensis*, registering this event until 45 days in the different tests performed). Studies carried out on *P. tarapacana* showed that the appearance of seedlings was concentrated between 30 and 40 days after sowing [15], whereas in the case of *P. australis*, the first seedlings were observed at 20 days after sowing and were increasing until 60 days [12].

The percentages of germination obtained were variable, in *P. neglecta* 10% was registered, in the case of *P. pacensis* 8% and for the species *P. incarum* 2%. There are several reports about the low germinative power of *Polylepis* species. For example, for *P. incana* (Peru) there is information that the percentage of germination varies between 2 and 15% [27]. In *P. tomentella* there is a 3.1% [28], and in *P. australis* there is a variation of 10–50% of germination percentage under experimental conditions [12]. In the study carried out by the Mining Company Doña Inés de Collahuasi SCM [15] it is mentioned that the percentage of germination in massive propagation in *P. tarapacana* fluctuated remarkably according to the origin of the seeds, in a range of 2–10%, with an average close to 4%. In *P. besseri*, the conclusion of the germinative stage occurred between 50 and 65 days, with a germination rate of 3–10% in experimental plots [1], whereas, in *P. racemosa*, the germinative success was of 2–15%, in a period of 60–70 days [25]. Therefore, the success rate of propagation will depend mainly on the amount of viable seeds and the adequate environmental conditions (soil, humidity and temperature) that are provided for germination [11]. On the other hand, it is important to take into account the distribution of the seed trees from which the samples were obtained, either in the city or in the field, an aspect that can respond to the low germination rate presented in the species under study. The crossing between neighboring plants can lead to a biparental depression of inbreeding, that is, a loss of biological effectiveness (vigor, viability, fecundity) due to the reduction of genetic variability due to homozygosity, which reduces the chances of survival of the species [23]. For this reason, it should be taken into account that the collection of seeds takes place in places where a broad dispersion flow can be evidenced that, in a certain way, guarantees greater genetic variability within that population.

5. Conclusions

Through this study, it was possible to generate basic reference information, providing information on the location, description and selection of seedlings of *Polylepis* in areas near or belonging to the Municipality of La Paz, to determine parameters related to the physical and physiological quality of seeds, and also the development of basic procedures for their initial cultivation in controlled conditions (greenhouse).

In this way, it is intended to contribute to the biological knowledge of the species examined, filling some gaps of information to understand their current status and plan the development of appropriate *in situ* and *ex situ* conservation strategies for the species under study through restoration or reforestation programs with seedlings obtained by seed.

Acknowledgements

We thank the Department of Postgraduate Research and Social Interaction-DIPGIS, for the financing granted to carry out this research through the IDH 2013-2014 UMSA competitive funds. We thank the researchers I. Gómez, A. Palabral and A. Domic, for the close collaboration with this initiative, which allowed to develop a synergic work of interdisciplinary cooperation; and Ms. R.I. Meneses, for allowing us the use of environments for the processing and storage in the National Herbarium of Bolivia of the botanical collections of the project.

Author details

Cecilia Vega Krstulovic*, Jorge Quezada Portugal, Paola Rocabado Koya, Gabriela Villegas Alvarado and Juan Carlos Bermejo Franco
Biotechnology Plant Unit of the Institute of Molecular Biology and Biotechnology, Universidad Mayor de San Andrés (U.M.S.A.), La Paz, Bolivia

*Address all correspondence to: kcecivegak@hotmail.com

References

[1] Hensen I. Estudios ecológicos y fenológicos sobre *Polylepis besseri* Hieron en la cordillera oriental Boliviana. Ecología en Bolivia. 1994;**23**:21-26

[2] Fjeldså J, Kessler M. Conservación de la Biodiversidad de los Bosques de *Polylepis* de Las Tierras Altas de Bolivia. Una Contribución al Manejo Sustentable en los Andes. DIVA Technical Report 11. Santa Cruz de la Sierra, Bolivia: Editorial FAN; 2004

[3] Kessler M. Present and potential distribution of *Polylepis* (Rosaceae) forests in Bolivia. In: Churchill SP, Balslev H, Foresro E, Luteyn JL, editors. Botánica Económica de los Andes Centrales. Biodiversity and Conservation of Neotropical Montane Forests. New York, EE.UU; 1995

[4] Bitter G. Revision der Gattung *Polylepis*. Botanische Jahrbücher für Systematik. 1911;**45**:564-656

[5] Simpson B. A revision of genus *Polylepis* (Rosaceae: Sanguisorbae). Smithsonian Contributions to Botany. 1979;**43**:1-62

[6] Kessler M, Schmidt-Lebuhn AN. Taxonomical and distributional notes on *Polylepis* (Rosaceae). Organisms, Diversity and Evolution. 2006;**6**:67-69

[7] Gareca EE, Hermy M, Fjeldså J, Honnay O. *Polylepis* woodland remnants as biodiversity islands in the Bolivian high Andes. Biodiversity and Conservation. 2010;**19**:3327-3346

[8] Kessler M. Bosques de *Polylepis*. In: Moraes M, Øllgaard B, Kvist LP, Borchsenius F, Balslev H, editors. Botánica Económica de los Andes Centrales. La Paz, Bolivia: Universidad Mayor de San Andrés; 2006. pp. 110-120

[9] Tarifa T, Yensen E. Mamíferos de los bosques de *Polylepis* de Bolivia. Revista Boliviana de Ecología y Conservación Ambiental. 2001;**9**:29-44

[10] Kessler M, Driesch P. Causas e historia de la destrucción de bosques altoandinos de Bolivia. Ecología en Bolivia. 1993;**21**:1-18

[11] Prettell J, Ocaña D, Jon R, Barahona E. Apuntes Sobre Algunas Especies Forestales Nativas de la Sierra Peruana. Lima, Perú: Proyecto FAO Holanda. Infor; 1985

[12] Renison D, Cingolani A. Experiencias en germinación y reproducción vegetativa aplicados a la reforestación con *Polylepis australis* (Rosaceae) en las sierras Grandes de Córdoba, Argentina. Ağrı. 1998;**15**:47-53

[13] Renison D, Hensen I, Cingolani AM. Anthropogenic soil degradation affects seed viability in *Polylepis australis* mountain forests of Central Argentina. Forest Ecology and Management. 2004;**196**:327-333

[14] MMAYA. Plan Nacional Para la conservación y Manejo Sostenible de los Bosques de *Polylepis* (Kewiña, Keñua y/o Lampaya) y Su Biodiversidad Asociada (Programas y Lineamientos Generales). La Paz, Bolivia: EDOBOL; 2012

[15] Compañía Minera Doña Inés de Collahuasi SCM. Queñoa árbol de Las Alturas. Chile: Santiago de Chile; 2011

[16] Arrázola S, Coronado I. *Polylepis pacensis* M. Kessler and Schmidt-Leb. Pp. 338-339, *Polylepis neglecta* M. Kessler. Pp. 535-536 en Libro Rojo de la Flora Amenazada de Bolivia, Volumen I. Zona Andina, Ministerio de Medio Ambiente y Agua, La Paz, Bolivia; 2012

[17] Arrázola S, Coronado I, Torrico L, Meneses RI, Navarro G, Ferreira W. *Polylepis incarum* (Bitter) M. Kessler and Scmidt-Leb. en Libro Rojo de la Flora

Amenazada de Bolivia, Volumen I. Zona Andina, Ministerio de Medio Ambiente y Agua, La Paz, Bolivia. 2012. pp. 44-45

[18] International Seed Testing Association (ISTA). International Rules for Seed Testing. Rules. Zurich, Suiza: Seed Science and Technology; 1993

[19] Danida, FAO. Guía Para la manipulación de Semillas Forestales. Roma, Italia: Centro de Semillas Forestales; 1991

[20] Ciencia y Tecnología de Semillas. Módulo Análisis de Semillas. In: Curso de Especialización en Ciencia y Tecnología de Semillas. Brasil: Universidad Federal de Pelotas RS; 2002

[21] Bautista A. Manual de Ensayos de Semillas Forestales. México: Secretaría de Medio Ambiente; 2012

[22] Serrada R. Apuntes de Repoblaciones Forestales. Madrid, España: FUCOVASA; 2000

[23] Seltmann P, Leyer I, Renison D, Hensen I. Variation of seed mass and its effects on germination in *Polylepis australis*: Implications for seed collection. New Forests. 2007;**33**:171-181

[24] CESA. Experiencias sobre reforestación en la sierra ecuatoriana, con especies nativas. Ecuador; 1991. p. 30. Available from: https://goo.gl/tYyL7U (consultada en: junio 2018)

[25] Reynel CR, León JG. Árboles y Arbustos Andinos Para agroforestería y conservación de Suelos. Perú: Tomo II. Las Especies. Proyecto FAO Holanda-DGFF; 1990

[26] BASFOR. Banco de Semillas Forestales. Cochabamba, Bolivia: Catálogo de semillas forestales; 2001

[27] Fjeldså J, Kessler M. Conserving the Biological Diversity of *Polylepis* Woodlands of the Highland of Peru and Bolivia. A Contribution to Sustainable Natural Resource Management in the Andes. Copenhagen, Denmark: NORDECO; 1996

[28] Cruz AS. Efecto de sustratos orgánicos en la reproducción vegetativa de la queñua (*P. incana*, H.B.K.) Rosoideae [tesis de Grado]. La Paz. Bolivia: Facultad de Agronomía, Universidad Mayor de San Andrés; 1999. pp. 90-92

Chapter 7

Phytotoxicity of *Plantago major* Extracts on Germination and Seedling Growth of Purslane (*Portulaca oleracea*)

Ahmed F. Al-obaidi

Abstract

Plantago major L. (Plantaginaceae family) has been used as herbal remedies for centuries in almost all over the world and in the treatment of a number of diseases. This study aims to assess the allelopathic potential of *Plantago major* extracts on the germination and early seedling growth of purslane. Total phenols, tannins, saponins, flavonoids, and alkaloids were determined in *P. major*. Furthermore, concentrations of 2.5, 5, 10, 20, and 40 mg.ml^{-1} of both alcoholic and aqueous extracts were prepared to study their phytotoxic effect on the germination and seedling growth of *Portulaca oleracea* weed. In our study, showing the germination of *P. oleracea* was completely inhibited (96.30 mg.ml^{-1}) under treatment of *P. major* methanolic extracts at 40 mg ml^{-1}. Moreover, both radicle and plumule were strongly inhibited (87.20 and 74.29 mg.ml^{-1}, respectively) under the same treatment. This could be attributed to the high content of bioactive constituents. Therefore, this species can be used in the method of biological control of weeds. In addition, further studies are required to identify and characterize the proper allelochemicals and demonstrate their modes of action.

Keywords: allelopathy, *Plantago major*, *Portulaca oleracea*, phytochemical

1. Introduction

At present there is a lot of emphasis on finding new methods to fight weeds, and concept of competition between plant species has been improved with that of plant allelopathy [1, 2]. Allelopathy involves the effects of one plant on another because of the chemicals it releases or the breakdown products of their metabolites [3]. There are some examples of plant toxins among the plant secondary compound classes of alkaloids, terpenes, and especially phenolics [4]. Phytotoxicity assays have been reported to be an important approach for identifying plants that are likely to be a source of vital herbivorous compounds [5, 6].

The allelopathic effects of crop plants or crop residues on weeds benefit farmers, which can cause significant economic losses [7]. There is competition for weed crops for moisture, nutrients, space, and light, which negatively affects crop yield [8]. It has been reported that the predominant species of weed allelochemicals stop crop production but sometimes also stimulate seed growth, germination, and crop production [9, 10].

Management methods that reduce the requirement for herbicides are needed to reduce adverse environmental impacts. Herbicides can cause crop injury [11]. Moreover, there is a keen interest in developing alternative methods of natural weed control in organically grown crops [12], as weed control remains one of the most significant agronomic challenges in the production of organic crops. Weed management is often the most troublesome technical problem to be solved in organic farming, especially in poorly competitive crops like vegetables [6, 13]. Cultivation and hand hoeing are common practices used in organically grown leek crops.

Portulaca oleracea L. (purslane) is a common troublesome weed worldwide. Despite being considered a poor competitor, it can quickly establish and easily regenerate by vegetative reproduction method [14]. *Plantago* is the largest genus within the Plantaginaceae family comprising approximately 275 annual and perennial species distributed all over the world [9]. *Plantago major* L. (*Plantago major ssp. major* L.) is a perennial plant that belongs to the Plantaginaceae family and is found in fields, lawns, and on the roadsides. It can become about 10–60 cm high, but the size varies a lot depending on the growth habitats. The leaves grow in rosettes, and they are ovate to elliptical with parallel venation (5–9) [15]. In Asia and Europe, the aerial parts of *P. major* is often used as herbal remedies in the treatment of a number of diseases related to the skin, respiratory and digestive organs, reproduction, and against infections [16].

Phytochemical investigation of the genus revealed the presence of polysaccharides, phenylpropanoid glycosides, alkaloids, triterpenes, flavonoids, and phenolic acids as the main bioactive compounds present in the aerial parts [17–20]. The aim of the present study was to evaluate the allelopathic potential of *Plantago major* extracts on the germination and early seedling growth of purslane.

2. Materials and methods

2.1 Plant material

Plantago major L. was collected from canal banks in Al Anbar city (Iraq) during their vegetative stage (February 2018). The identification of species was done according to Boulos [15]. The plant material was handily cleaned, washed several times with distilled water to remove dust and other residues, dried in room temperature in shaded place for several days till complete dryness and ground into powder, and then preserved in well-stopped bottles [21].

2.2 Phytochemical analysis

Plantago major was collected and prepared as previously mentioned. Total phenolics, flavonoids, and alkaloids were estimated using spectrophotometric techniques adapted by Harborne [22], Sadasivam and Manickam [23], and Boham and Kocipai-Abyazan [24], respectively. Tannins were determined according to Van-Buren and Robinson [25], while saponin content was estimated by the method adopted by Obadoni and Ochuko [26].

2.3 Allelopathy bioassay weed seed source

The seeds of *Portulaca oleracea* were collected from cultivated land from Al Anbar, Iraq. Seeds were sterilized by 0.3% sodium hypochlorite for 3 minutes, washed several times by distilled water, dried at room temperature for 7 days, and reserved in paper bag until further use [27, 28].

2.4 Preparation of extracts

For bioassay tests, aqueous and methanol extracts were prepared to obtain various concentrations of 2.5, 5, 10, 20, and 40 mg.ml^{-1} (w/v). The solutions were filtered through double layers of muslin cloth followed by Whatman No. 1 filter paper. The pH of the mixtures was adjusted to 7 with 1 M HCl, and then mixtures were stored in a refrigerator at 4°C until further use [29].

2.5 Germination bioassay

For germination experiment, 25 seeds were placed in each filter paper in addition to 10 ml of tested extract for each Petri dish (90 mm diameter). The control treatment was designed with distilled water. Germinated seeds were counted daily starting from the first day of treatment. The design of the experiment was randomized complete block with three replicates. The experiment was repeated three times, and the inhibition percentage was calculated.

2.6 Seedling growth bioassay

The seeds of *Portulaca oleracea* were germinated in the dark at room temperature for 2 days. Twenty-five germinated seeds were placed in Petri dishes lined with two layers of filter paper (Whatman No. 1), and 10 ml of different extracts (2.5, 5, 10, 20, and 40 mg.ml^{-1}) were added. Moreover, a control treatment was designed with distilled water. The design of the experiment was randomized complete block with three replicates. The experiment was repeated twice, the radicle and plumule lengths of seedlings were measured on the tenth day, and growth inhibition for radicle and plumule lengths were calculated.

3. Results and discussion

3.1 Phytochemical constituents

Several phytotoxic substances causing germination and/or growth inhibitions have been isolated from plant tissues [30, 31]. The phytochemical constituents of *Plantago major* are presented in **Table 1**. *Plantago major* contained high contents of phenolics (132.2 mg/g dry weight) and tannins (28.7 mg/g dry weight), while it contained relatively alkaloids (10.6 mg/g dry weight), saponins (15.8 mg/g dry weight), and flavonoids (14.8 mg/g dry weight).

These results are supported with the study of Kolak et al. [32] and Miser-Salihoglu et al. [33]. In addition, this results relatively comparable to those reported in *Senecio glaucus* as described by El-Amier et al. [34] with the exception of phenols less, but higher than those reported by Kobeasy et al. [35] on same species and

Plant species	Active organic compounds (mg.g^{-1} dry weight)				
	Phenolics	Tannins	Alkaloids	Flavonoids	Saponins
Plantago major	132.2 ± 2.35	28.7 ± 0.89	10.6 ± 0.05	14.8 ± 0.21	15.8 ± 0.06

Table 1.
Concentrations of the active organic compounds estimated in Plantago major.

El-Amier et al. [36] on *Euphorbia terracina* as well as El-Amier and Abdullah [37] on some wild plants (*Calligonum polygonoides*, *Cakile maritima*, and *Senecio glaucus*).

3.2 Allelopathic effect of *P. major* extracts on *P. oleracea* germination

Allelopathy is a phenomenon by which some plants affect the others, either positively or negatively, by exuding chemicals [38]. In the present study, the allelopathic effect of shoot extracts (aqueous and methanol) on the germination percentage of *Portulaca oleracea* at 4 DAT was shown in **Figure 1**. It is observed from the figure that the methanolic extract of *Plantago major* exhibited higher germination inhibition of *Portulaca oleracea* than the aqueous extract. This could be attributed to the methanol polarity that has the ability to extract a wide variety of active components compared to water [39]. The degree of inhibition was significantly increased in a concentration-dependent manner. The aqueous extract of *P. major* at 40 mg ml^{-1} inhibited the germination of *P. oleracea* by about 30.24%, while the lowest concentration (2.5 mg ml^{-1}) inhibited the germination by 4.60%. On the other hand, *P. major* methanolic extract showed a highest inhibition of germination at 40 mg ml^{-1}, while at 2.5 mg ml^{-1}, it exhibited lowest inhibition percentage (20.37%).

Many plant species showed inhibitory effects on *P. oleracea* germination such as *Medicago sativa* and *Vicia cracca* [40], *Salvia macrochlamys* [41], wheat, and rye straw [42]. Aqueous extract of some plant species may contain some toxic substances [43]. These substances probably inhibit the germination and seedling growth of other plant species [44], which was due to their interference with indole acetic acid metabolism, or synthesis of protein and ion uptake by the plants [45].

3.3 Allelopathic effect of *P. major* extracts on *P. oleracea* seedling growth

Allelopathy offers potential for biorational weed control through the production and release of allelochemics from leaves, flowers, seeds, stems, and roots of living or decomposing plant materials. Under appropriate conditions, allelochemics often exhibit selectivity, similar to synthetic herbicides [46].

The allelopathic effect of both aqueous and methanolic extracts on *Portulaca oleracea* radicle growth after 10 days of treatment revealed that there was significant variation between different extracts. However, the degree of inhibition significantly

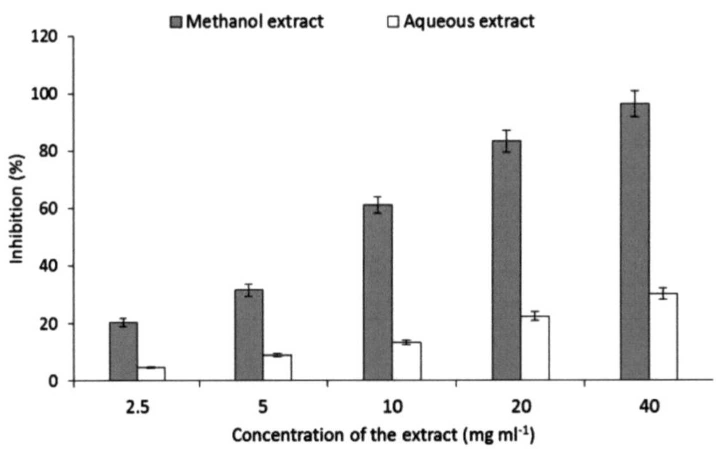

Figure 1.
The allelopathic effect of both aqueous and methanolic Plantago major extracts on the germination inhibition percentage (mean value) with the error bars of Portulaca oleracea 10 days after treatment.

increased in a dose-dependent manner (**Figure 2**). The aqueous extract of *P. major* showed 52.34% at 40 mg ml^{-1}, while it showed the lowest inhibition percentage of radicle growth (3.5%) at 2.5 mg ml^{-1} (**Figure 2**). On the other side, the methanolic extracts from *P. major* at 40 mg ml^{-1} inhibited the radicle growth of *Portulaca oleracea* by 87.20%, while at the lowest concentration (2.5 mg ml^{-1}), *P. major* extract showed the lowest inhibition percentage (19.21%) of radicle growth (**Figure 2**).

The phytotoxic effect of both methanolic and aqueous extracts from the studied *Plantago* species on *Portulaca oleracea* plumule growth revealed slight significant variation between two extracts. However, there was a very large difference between different concentrations (**Figure 3**). The aqueous extract from *P. major* showed the highest inhibition percentage of *Portulaca oleracea* plumule growth (48.69%) at 40 mg ml^{-1}, while at 2.5 mg ml^{-1}, *P. major* extract inhibited the plumule growth by 4.11%. On the other hand, the methanolic extract of *P. major* exhibited high inhibition (74.29%) of *P. oleracea* plumule growth at 40 mg ml^{-1}. The lowest concentration (2.5 mg ml^{-1}) of *P. major* extract inhibited the plumule growth by 60.95% (**Figure 3**). Phytochemical investigation of the genus revealed the presence of

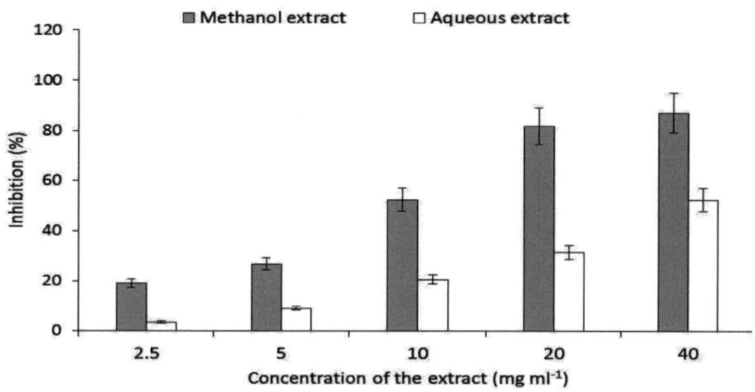

Figure 2.
The allelopathic effect of both aqueous and methanolic Plantago major extracts on the radicle growth inhibition percentage (mean value) with the error bars of Portulaca oleracea 10 days after treatment.

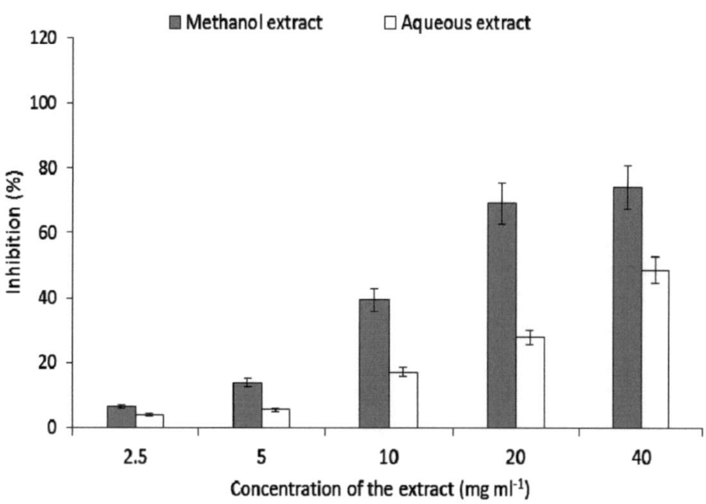

Figure 3.
The allelopathic effect of both aqueous and methanolic Plantago major extracts on the plumule growth inhibition percentage (mean value) with the error bars of Portulaca oleracea 10 days after treatment.

phenylpropanoid glycosides, alkaloids, triterpenes, flavonoids, and phenolic acids as the main bioactive compounds present in the aerial parts [17, 19, 20].

The allelopathic effect of *P. major* could be attributed to several bioactive compounds that act in a synergistic manner or to compounds which regulate one another such as flavonoid, phenolic acids, saponin, alkaloids and tannins. *Plantago* species was reported to contain several bioactive secondary metabolites such as vanillic acid, iridoid glycoside (aucubin), caffeic acid derivatives, chlorogenic acid, ferulic acid, *p*-coumaric acid, and triterpenes (oleanolic acid, ursolic acid) [16, 47, 48]. Many of these compounds were reported as allelochemicals [49]. Generally, the reduction in the seedling growth of *P. oleracea* in this study may be attributed to reduction in cell division of the seedlings, altering the ultrastructure of the cells as well as leading to the alteration of the ion uptake, water balance, phytohormone balance, photosynthesis, respiration, and inactivate several enzymes [50, 51].

4. Conclusion

In conclusion, the aim of this study was to assess the allelopathic potential of *Plantago major* extracts on the germination and early seedling growth of purslane. In our study, showing the germination of *Portulaca oleracea* was completely inhibited under treatment of *P. major* methanolic extracts at 40 mg ml^{-1}. Moreover, both radicle and plumule were strongly inhibited under the same treatment. This could be attributed to the high content of bioactive constituents. Therefore, this species can be used in the method of biological control of weeds. In addition, further studies are required to identify and characterize the proper allelochemicals and demonstrate their modes of action.

Author details

Ahmed F. Al-obaidi
Horticulture Science Department, Collage of Basic Education, Anbar University, Haditha, Iraq

*Address all correspondence to: drahmed628@gmail.com

References

[1] Razzaq ZA, Cheema K, Jabran K, Farooq M, Khaliq A, Haider G, et al. Weed management in wheat through combination of allelopathic water extract with reduced doses of herbicides. Pakistan Journal of Weed Science Research. 2010;**16**(3):247-256

[2] Wang H, Zhou Y, Chen Y, Wang Q, Jiang L, Luo Y. Allelopathic potential of invasive *Plantago virginica* on four lawn species. PLoS One. 2015;**10**(4):e0125433. DOI: 10.1371/journal.pone.0125433

[3] Willis RJ. Terminology and trends in allelopathy. Allelopathy Journal. 1994;**1**(1):6-28

[4] Aasifa G. Allelopathic effect of aqueous extracts of different part of *Eclipta alba* (L.) Hassk. On some crop and weed plants. Journal of Agricultural Extension and Rural Development. 2014;**6**:55-60

[5] Scognamiglio M, D'Abrosca B, Esposito A, Pacifico S, Monaco P, Fiorentin A. Plant growth inhibitors: Allelopathic role or phytotoxic effects. Focus on Mediterranean biomes. Phytochemistry Reviews. 2013;**12**(4):803-830

[6] Trezzi MM, Vidal RA, Balbinot AA, Bittencourt HV, Souza A. Allelopathy: Driving mechanisms governing its activity in agriculture. Journal of Plant Interaction. 2016;**11**:53-60

[7] Reinhardt CF, Meissner R, Labuschagne N. Allelopathic interaction of *Chenopodium album* L. and certain crop species. South African Journal of Plant and Soil. 1994;**11**:45-49

[8] Kadioglu I, Yanar Y, Asav U. Allelopathic effects of weed extracts against seed germination of some plants. Journal of Environmental Biology. 2005;**26**(2):169-173

[9] Goncalves S, Romano A. The medicinal potential of plants from the genus *Plantago* (Plantaginaceae). Industrial Crops and Products. 2016;**83**:213-226

[10] Narwal SS. Allelopathy in Crop Production. Jodhpur, India: Scientific Publisher; 1994. p. 288

[11] Bilalis D, Efthimiadis P, Katagiannis G. The phytotoxicity of various graminicides in durum wheat in Greece. Journal of Agronomy and Crop Science. 2001;**187**:121-126

[12] Bilalis D, Papastylianou P, Konstantas A, Patsiali S, Karkanis A, Efthimiadou A. Weed-suppressive effects of maize–legume intercropping in organic farming. International. Journal of Pest Management. 2010;**56**:173-181

[13] Peruzzi A, Ginanni M, Fontanelli M, Raffaelli M, Bàrberi P. Innovative strategies for on-farm weed management in organic carrot. Renewable Agriculture and Food Systems. 2007;**22**:246-259

[14] Mohamed AI, Hussein AS. Chemical-composition of purslane (*Portulaca oleracea*). Plant Foods for Human Nutrition. 1994;**45**(1):1-9

[15] Boulos L. Flora of Egypt. In: Verbenaceae–Compositae. Vol. 3. Cairo, Egypt: AL-Hadara Publishing; 2002

[16] Samuelsen AB. The traditional uses, chemical constituents and biological activities of *Plantago major* L. A review. Journal of Ethnopharmacology. 2000;**71**:1-21

[17] Haddadian K, Haddadian K, Zahmatkash M. A review of Plantago plant. Indian Journal of Traditional Knowledge. 2014;**13**(4):681-685

[18] Ronsted N, Gobel E, Franzyk H, Rosendal JS, Olsen CE. Chemotaxonomy of *Plantago*. Iridoid glucosides and caffeoyl phenylethanoid glycosides. Phytochemistry. 2000;**55**:337-348

[19] Tarvainen M, Suomela JP, Kallio H, Yang B. Triterpene acids in *Plantago major*: Identification, quantification and comparison of different extraction methods. Chromatographia. 2010;**71**:279-284

[20] Taskova R, Evstatieva LJ, Handjieva N, Popov S. Iridoid patterns of genus *Plantago* L. and their systematic significance. Zeitschrift für Naturforschung. 2002;**57c**:42-50

[21] AOAC. Official Methods of Analysis. 15th ed. Arlington, Virginia, USA: Association of Official Analytical Chemists; 1990

[22] Harborne JB. Phytochemical Methods. London: Chapman and Hall, Ltd; 1973. pp. 49-188

[23] Sadasivam S, Manickam A. Biochemical Methods, New Age International Limited. 3rd ed. New Delhi; 2008

[24] Boham BA, Kocipai-Abyazan R. Flavonoids condensed tannin from leaves of Hawaiian *Vaccinium reticulatum* and *V. calycinum*. Pacific Science. 1994;**48**:458-463

[25] Van-Buren JP, Robinson WB. Formation of complexes between protein and tannic acid. Journal of Agricultural and Food Chemistry. 1969;**17**:772-777

[26] Obadoni BO, Ochuko PO. Phytochemical studies and comparative efficacy of the crude extracts of some homeostatic plants in Edo and delta states of Nigeria. Global Journal of Pure and Applied Sciences. 2001;**8**:203-208

[27] Sampietro DA, Catalan CAN, Vattuone MA. Isolation, Identification and Characterization of Allelochemicals Natural Products. USA Science Publishers: Enfield, NH; 2009. p. 2009

[28] Uremis I, Arslan M, Uludag A. Allelopathic effects of some *Brassica* species on germination and growth of cut leaf ground cherry (*Physalis angulata* L.). Journal of Biological Science. 2005;**5**:661-665

[29] Rice EL. Allelopathic effect of *Andropogon virginicus* and its persistence in old field. American Journal of Botany. 1972;**59**:752-755

[30] Soyler D, Canýhoþ E, Temel N, Hajyzadeh M. Determination of chemical fungicide against soil borne fungal diseases of capers (*Capparis ovata* Desf. var. *herbacea*) during early stages. Pakistan Journal of Agriculture Science. 2012;**49**:345-348

[31] Turk MA, Tawaha AM. Allelopathic effect of black mustard (*Brassica nigra* L.) on germination and growth of wild oat (*Avena fatua* L.). Crop Protection. 2003;**22**:673-677

[32] Kolak U, Boga M, Akalin Urusak E, Ulubelen A. Constituents of *Plantago major* subsp. *intermedia* with antioxidant and anticholinesterase capacities. Turkish Journal of Chemistry. 2011;**35**:637-645

[33] Miser-Salihoglu E, Akaydin G, Caliskan-Can E, Yardim- Akaydin S. Evaluation of antioxidant activity of various herbal folk medicines. Journal of Nutrition and Food Sciences. 2013;**3**(5):1-9

[34] El-Amier YA, Abdelghan AM, Zaid AA. Green synthesis and antimicrobial activity of *Senecio glaucus*-mediated silver nanoparticles. Research Journal of Pharmaceutical, Biological and Chemical. 2014;**5**:631-642

[35] Kobeasy MI, Abdel-Fatah M, Abd El-Salam SM, Mohamed ZM. Biochemical studies on *Plantago major* L. International Journal of Biodiversity and Conservation. 2011;**3**:83-91

[36] El-Amier YA, Al-Hadithy ON, Abdulhadi HL, Fayed EM. Evaluation of antioxidant and antimicrobial activities of *Euphorbia terracina* L. from deltaic Mediterranean coast, Egypt. Journal of Natural Products and Resources. 2016;**2**(2):83-85

[37] El-Amier YA, Abdullah TJ. Allelopathic effect of four wild species on germination and seedling growth of *Echinochloa crus-galli* (L.) P. Beauv. International Journal of Advanced Research. 2014;**2**(9):287-294

[38] Chon SU, Kin Y, Kee JC. Herbicidal potential and quantification of causative allelochemicals from several compositae weeds. Weed Research. 2003;**43**:444-450

[39] Oskoueian E, Abdullah N, Ahmad S, Saad WZ, Omar AR, Ho YW. Bioactive compounds and biological activities of *Jatropha curcas* L. kernel meal extract. International Journal of Molecular Sciences. 2011;**12**(9):5955-5970

[40] Koloren O. Allelopathic effects of *Medicago sativa* L. and *Vicia cracca* L. leaf and root extracts on weeds. Pakistan Journal of Biological Sciences. 2007;**10**:1639-1642

[41] Erez ME, Fidan M. Allelopathic effects of sage (*Salvia macrochlamys*) extract on germination of *Portulaca oleracea* seeds. Allelopathy Journal. 2015;**35**(2):285-296

[42] Boz Ö. Allelopathic effects of wheat and rye straw on some weeds and crops. Asian Journal of Plant Sciences. 2003;**2**(10):772-778

[43] Habib SA and Abdul- Rehman AA. Evaluation of some weed extracts against dodder on alfalfa (*Medicago sativa*). Journal of Chemical Ecology. 1988;**14**:443-452

[44] Al-Charchafchi FMR, Redha FMJ, Kamel WM. Dormancy of *Artemisia herba alba* seeds in relation to endogenous chemical constituents. Journal of Biological Sciences Research, Baghdad/Iraq. 1987;**18**:1-12

[45] Hussain F, Khan TW. Allelopathic effects of Pakistani weed *Cynodon dactylon* L. Journal of Weed Science Research. 1988;**1**:8-17

[46] Weston LA. Utilization of allelopathy for weed management in agroecosystems. Agronomy Journal. 1996;**88**:860-866

[47] Chiang LC, Chiang W, Chang MY, Ng LT, Lin CC. Antiviral activity of *Plantago major* extracts and related compounds *in vitro*. Antiviral Research. 2002;**55**:53-62

[48] Long C, Moulis C, Stanislas E, Fouraste I. L'aucuboside et le catalpol dans les feuilles de *Plantago lanceolata* L., *Plantago major* L. et *Plantago media* L. Journal de Pharmacie de Belgique. 1995;**50**:484-488

[49] Cheema ZA, Farooq M, Wahid A. Allelopathy: Current Trends and Future Applications. Berlin: Springer-Verlag Berlin Heidelberg; 2013. pp. 113-143

[50] Fahmy GM, Al-Sawaf NA, Turki H, Ali HI. Allelopathic potential of *Pluchea dioscoridis* (L.) DC. Journal of Applied Sciences Research. 2012;**8**:3129-3142

[51] Li Z, Wang Q, Ruan X, Pan C, Jiang D. Phenolics and plant allelopathy. Molecules. 2010;**15**:8933-8952